AF841969

The Grand Ethiopian Renaissance Dam, its Impact on Egyptian Agriculture and the Potential for Alleviating Water Scarcity

Environment & Policy

VOLUME 55

More information about this series at http://www.springer.com/series/5921

Youssef M. Hamada

The Grand Ethiopian Renaissance Dam, its Impact on Egyptian Agriculture and the Potential for Alleviating Water Scarcity

Springer

Youssef M. Hamada
Agriculture Research Center
Agricultural Economics Research Institute
Cairo
Egypt

ISSN 1383-5130 ISSN 2215-0110 (electronic)
Environment & Policy
ISBN 978-3-319-54438-0 ISBN 978-3-319-54439-7 (eBook)
DOI 10.1007/978-3-319-54439-7

Library of Congress Control Number: 2017932429

© Springer International Publishing AG 2017
This work is subject to copyright. All rights are reserved by the Publisher, whether the whole or part of the material is concerned, specifically the rights of translation, reprinting, reuse of illustrations, recitation, broadcasting, reproduction on microfilms or in any other physical way, and transmission or information storage and retrieval, electronic adaptation, computer software, or by similar or dissimilar methodology now known or hereafter developed.
The use of general descriptive names, registered names, trademarks, service marks, etc. in this publication does not imply, even in the absence of a specific statement, that such names are exempt from the relevant protective laws and regulations and therefore free for general use.
The publisher, the authors and the editors are safe to assume that the advice and information in this book are believed to be true and accurate at the date of publication. Neither the publisher nor the authors or the editors give a warranty, express or implied, with respect to the material contained herein or for any errors or omissions that may have been made. The publisher remains neutral with regard to jurisdictional claims in published maps and institutional affiliations.

Printed on acid-free paper

This Springer imprint is published by Springer Nature
The registered company is Springer International Publishing AG
The registered company address is: Gewerbestrasse 11, 6330 Cham, Switzerland

Acknowledgements

I acknowledge the assistance of Fritz Schmuhl and thank the reviewers for advice on and help with this manuscript. This work is dedicated to my sister, my brothers, and in memory of my parents. Errors and omissions remain the responsibility of the author.

Prof. Youssef M. Hamada

Contents

About the Author

Prof. Youssef M. Hamada is a Professor of Agricultural Economy, Agricultural Economics Research Institute, Agricultural Research Center, Egypt. He has been a researcher in Central Laboratory for Design and Statistics Analysis Research, Agriculture Research Centre. He received his B.Sc. (1983) in Agriculture Economics, his M.Sc. (1995) in economics of bread industry in A.R.E from Al-Azhar University, and his Ph.D. (2001) in the possibilities of the bread-industry development in A.R.E from the University of Al-Azhar, Egypt. He received his Diploma in Translation Arabic—English from Translation Center, American University, Egypt. He was awarded training from ECARDA, Aleppo, Syria, 2005 (training course for building data bases and electronic Web sites on the Internet) and International Training Workshop on Economic Development Policies and Practices for Desertification and Dry Land for Affected African Countries from Chinese Academy of Forestry, Beijing, China (2008). He has published more than 30 papers in local and international journals as well as book chapters.

List of Figures

List of Tables

Chapter 1
Introduction

Abstract This chapter provides a brief overview of background of the book, purpose of the book, scope of the book, the Nile River, the Nile River Basin and water security, dams and barrages on river channels, effects of Dams and Barrages on river channels, the Legal Aspects of the Nile River, agriculture and irrigation in Nile Basin, agriculture and irrigation in Egypt, and efficiency use of irrigation water.

Keywords Purpose of the Book · The Nile River · The Nile River Basin and Water Security · Dams and Barrages on River Channels · Effects of Dams and Barrages on River Channels · The Legal Aspects of the Nile River · Agriculture and Irrigation in Nile Basin · Agriculture and Irrigation in Egypt · Efficiency Use of Irrigation Water

The key decisions are not about dams as such, but about options for water and energy development. They relate directly to one of the greatest challenges facing the world in this new century—the need to rethink the management of freshwater resources. A number of global initiatives and reports have documented the dramatic impact of human-induced water withdrawals from the world's lakes, rivers, and ground aquifers. Total annual freshwater withdrawals today are estimated at 3800 km^3—twice as much as just 50 years ago (WCD 2000).

The unfolding scenario for water use in many parts of the world is one of increasing concerns about access, equity, and the response to growing needs. This affects relations:

- within and between nations
- between rural and urban populations
- between upstream and downstream interests
- between agricultural, industrial, and domestic sectors
- between human needs and the requirements of a healthy environment.

The challenge is not to mobilize so as to compete successfully, but to cooperate in reconciling competing needs. It is to find ways of sharing water resources equitably and sustainably—ways that meet the needs of all people as well as those of the

© Springer International Publishing AG 2017

Y.M. Hamada, *The Grand Ethiopian Renaissance Dam, its Impact on Egyptian Agriculture and the Potential for Alleviating Water Scarcity*,
Environment & Policy 55, DOI 10.1007/978-3-319-54439-7_1

environment and economic development. These needs are all intertwined, and our challenge is to resolve competing interests collectively. Achieving equitable and sustainable solutions will be to the ultimate benefit of all. The imperative to supply growing populations and economies with water in a context of depleting groundwater resources, declining water quality, and increasingly severe limits to surface water extraction has brought sustainable water resources management to the top of the global development agenda. Although increasing competition for water suggests an expanding scope for conflict, it also provides an incentive for new forms of cooperation and innovation. Dire scenarios for water demand must not overshadow the fact that development paths that meet and manage the demand exist and are available for us to choose. History demonstrates that the path of cooperation has more often been followed than the path of conflict. The same must be true for our future (WCD 2000).

Background

During the twentieth century, large dams emerged as one of the most significant and visible tools for the management of water resources. The more than 45,000 large dams around the world have played an important role in helping communities and economies harness water resources for food production, energy generation, flood control, and domestic use. Current estimates suggest that some 30–40% of irrigated land worldwide now relies on dams and that dams generate 19% of world electricity. From the 1930s to the 1970s, the construction of large dams became—in the eyes of many—synonymous with development and economic progress. Viewed as symbols of modernization and humanity's ability to harness nature, dam construction accelerated dramatically. This trend peaked in the 1970s, when on average two or three large dams were commissioned each day somewhere in the world. While the immediate benefits were widely believed sufficient to justify the enormous investments made—total investment in large dams worldwide is estimated at more than $2 trillion—secondary and tertiary benefits were also often cited. These included food security considerations, local employment and skills development, rural electrification and the expansion of physical and social infrastructure such as roads and schools. The benefits were regarded as self-evident, while the construction and operational costs tended to be limited to economic and financial considerations that justified dams as a highly competitive option (WCD 2000).

Purpose of the Book

The objective of this book is to expand on the knowledge of current debates between Egypt and other eleven riparian states: Kenya, Sudan, South Sudan, Burundi, Eritrea (observer status), Republic of Congo, Rwanda, Uganda, Tanzania, Ethiopia states of River Nile on the issues of water security, and equitable, sustainable usage and management of the Nile River waters. It examines the reasons behind Ethiopia's bold approach in challenging the asymmetric power relations and looks into Egypt's unilateral approach to the usage of the Nile waters. The book also evaluates the role of new factors within and outside Ethiopia, considering the Nile Water System for universal water resource development and management (water being a public good at the local, regional, and international system levels). Either this book is concerned specifically with the response of river channels to a reduction in the supply of these sediments by dams and gravel mining. Both this book is concerned on the existing debates of "water security in the Nile River Basin" and the paradigm shifts of long-term sustainable, manageable, and equitable water-sharing arrangements within Ethiopia and within all the riparian states.

The scope of the book is to investigate Ethiopia's emerging influence in the Nile Basin and the country's use of various counter-hydro hegemonic strategies vis-à-vis Egypt and Sudan. It also evaluates the success of these strategies, in terms of changing Egypt's hydro-hegemonic toward a more cooperative arrangement. Moreover, this book seeks to find out why the adopted strategies are important that have been used are important to the highly contested and suggested multilateral "water-sharing" arrangements for "benefit-sharing." **Halla Qadami** defines benefit sharing as moving from the allocation of physical quantity of water to the economic, social, political, and environmental values it brings to all riparian states (Qadami 2008).

What is the legal status of the Nile water treaties described above—or more specifically, is the international legal regime established over the Nile through treaties concluded between Great Britain and Egypt with other powers still operational and binding on Nile Basin states? The answer to this question is fundamental to the issue of rights and obligations over Nile waters.

If the Nile Waters Treaties are valid and binding, they legitimize the legal order of the colonial period that gave Egypt preeminence in the control of the Nile and developments in the basin. This would be a severe constraint on the development efforts and opportunities of upper riparian states. But if the Nile Waters treaties are not binding, then the control and utilization of Nile waters are regulated by the principles of customary international water law. It would also mean that the Nile is in search of a new legal regime in the form of a basin-wide agreement. This would provide plenty of room for negotiation and bargaining as among the riparian states. It could help develop a utilization regime that is more sustainable and equitable (Okoth-Owiro 2004).

Scope of the Book

The River Nile is one of the longest rivers in the world (Martinon 2010). It flows for 6600 km, traversing more than 30° of latitude (Mason 2004). According to **Mason**, "The Nile Basin covers an area of 3.1 million km^2, of which 1% is urban, 2% are covered by forest, 3% by wetlands, 3% by open water, 4% by shrub, 5% by irrigated cropland, 10% by cropland, 30% by desert/semi-desert and 42% by grassland" (Mason 2004). Within the Nile Basin area approximately 160 million people depend on the River for their livelihoods (Kameri-Mbote 2007). It also provides life sustenance to human needs and to socio-economic activities for almost 500 million peoples (as of 2012 census) across the 11 riparian states (The Nile River 2010). It flows from the south (upstream states) to the north (downstream states) for 6825 km (Martinon 2010). Within all the riparian states, the Nile River faces present-day threats such as constant droughts, water pollution from land-use activities, and population growth (Kameri-Mbote 2007). This means that the Nile waters are "international" and they require the immediate attention of sustainable and manageable multilateral water-sharing arrangements.

The Nile River

The Nile, which is the only drainage outlet from Lake Victoria, is one of the longest rivers in the world. Its total length together with those of its tributaries is about 3,030,300 km. The catchment area of the Nile totals some 2,900,000 km^2, representing about one-tenth of the surface area of the entire African Continent (Okoth-Owiro 2004).

The Nile measures some 5611 km from its White Nile source in Lake Victoria (East Africa) and some 4588 km form the Blue Nile source in Lake Tsana (Ethiopia). Thus, the river system originates from two distinct geographical zones (Okoth-Owiro 2004).

One subsystem, with the White Nile as its main artery, originates in the equatorial lakes of East and Central Africa, the most important of which is Lake Victoria, and in the Bahr-el-Ghazal water system—a vast lagoon formed by the convergence of a number of streams rising to the East and North of the Nile-Congo divide (Okoth-Owiro 2004).

The other subsystem consists of the Blue Nile and its tributaries, the Atbara and the Sobat. It originates from the Ethiopian Plateau (Okoth-Owiro 2004).

The Nile is made up of three main tributaries. These are the White Nile, the Blue Nile, and the Atbara. The White Nile rises from its source in the highlands of Rwanda and Burundi and flows into Lake Victoria. It leaves Lake Victoria at its northern shore near the Ugandan town of Jinja, through a swampy stretch around Lake Kyoga in Central Uganda and then heads north toward Lake Albert (Okoth-Owiro 2004).

Lake Albert receives a good amount of water from the Semliki River, which has its source in the Congo and empties first into Lake Edward, where it receives additional water from tributaries coming from the Ruwenzori Mountains on its way to Lake Albert. From Lake Albert, the White Nile flows north into Southern Sudan (Kasimbaji 1998).

Lake Victoria, Edward, and Albert are the natural reservoirs, which collect and store great quantities of water from the high rainfall regions of eastern Equatorial Africa and maintain a permanent flow down the White Nile with relatively small seasonal fluctuations (Okoth-Owiro 2004).

In Southern Sudan, near the capital city of Khartoum, the White Nile meets the Blue Nile, which drains Lake Tsana in the Ethiopian Highlands. The two flow together to just north of Khartoum where some 108 km downstream, they are joined by the Atbara, the last important river in the Nile system, whose source is in Eritrea. The river then flows north through Lake Nasser and the Aswan Dam before splitting into major distributaries, the Rosetta and Damietta, just north of Cairo. These distributaries flow into the Mediterranean Sea (Okidi 1982).

According to Godana (1985), the average annual flow of the Nile is 84 milliards of cubic meters as measured in Aswan. Of this total, Batstone (1959) estimates that, 84% is contributed by Ethiopia and only 16% come from the Lake Plateau of Central Africa. A similar distribution pattern is given by Godana (1985) who asserts that 85% of the flow of the Nile originates from the Ethiopian plateau, whereas only 15 comes from the East African source areas. It is, however, important to note that the statistics of the flow of the Nile are a complex matter, which the above estimates tend to over simplify.

The Nile River Basin and Water Security

Water has always been a highly contentious issue in regional and international affairs, because of its importance for political and socio-economic stability of nations. Without access to water, the maximization and protection of human and economic development is difficult, if not impossible. The management of this scarce resource in equitable and sustainable terms is currently an important debate within the Nile Basin-wide region, addressing the issues of poverty, population growth, soil erosion, environmental degradation, and climate change. The Nile River waters are highly important to the evolving conflict between eleven riparian states: Kenya, Sudan, South Sudan, Burundi, Eritrea (observer status), Republic of Congo, Rwanda, Uganda, Tanzania, Ethiopia, and Egypt.

Egypt and Sudan, perceived as hydro-hegemons, continue to exhibit a growing concern over Ethiopia's move to divert some of the waters of the Blue Nile River for the purpose of hydroelectricity. This means that both countries' favorable colonial-legal agreements, signed between Egypt, Sudan, and Britain in the allocation of the Nile Waters, could be in jeopardy (Cascao 2011). Especially, Egypt's principal life sustenance depends on the Nile River (Hassan and Al Rasheedy 2011), and one of the Egypt's geographical challenges is the inhabitable desert and nearly zero rainfall. **Daniel Kandie** stresses that the Nile then drains into Egypt—a country where there is practically no rain, and 86% of the land is classified as very arid, and the rest as arid (Kandie 1999). Water, for Egypt and for other riparian states, is thus often a part of discourses and practices of "securitization" (Abrahamsen 2005) and "militarization" guided by fear, mistrust, and "zero-sum" politics. Paradoxically, what is an issue of national "security" (Buzan et al. 1998) and life and death for Egypt, it also so for all other riparian states. **Okbazghi Yohannes** places the recently growing concerns into perspective by arguing that now all Nile countries define water resources in terms of national security, hence the elevation of water resources to the status of high politics (Yohannes 2013).

This part of the book focuses on the existing debates of "water security in the Nile River Basin" and the paradigm shifts of long-term sustainable, manageable, and equitable water-sharing arrangements within Ethiopia and within all the riparian states. Ideally, this would mean the emergence of regional economic cooperation and integration. I have chosen this topic to draw attention to the fundamental changes taking place within Africa, and more specifically within Ethiopia. Moreover, the topic could not be more timely appropriate, since the issue (as of 2013) has reached a critical moment of trans-boundary relations on access to the Nile Waters (Tadesse 2013). Some of the key changes seen in the last decade are the constant increase of population and economic growth in all riparian states, particularly in Ethiopia.

In recent years, Ethiopia has come out strong against the asymmetric power relations governing access and utilization of the Blue Nile waters within its own territory. Ethiopia's unilateral action for hydraulic development on the Blue Nile River is supported by the Berlin, Helsinki, and International Law Commission (ILA 2013) rules on "equitable" and "no-harm" water usage (Tafesse 2011). The Nile waters, thus, are not only about "rights" to access and utilization, but also about managing the scarce resource equitably and sustainably with all riparian states involved.

Ethiopia is building the momentum to create a sense of urgency of the issue of unequal water usage distribution and management against Egypt and Sudan, using unilateral, bilateral, and multilateral strategies. As Egypt continues to stress its dependence on the Nile, it is considered a matter of national security for Egypt not to accept a genuine basin-wide cooperative legal framework (Martinon 2010). Contrary to this old and static view, Ethiopia has proceeded unilaterally to utilize some of the Blue Nile waters for domestic and regional-wide electricity production. One of the many reasons for Ethiopia's initial move is that many of Ethiopian rivers, the Abbay (Blue Nile), Baro-Akobo (Sobat), and Tekeze (Atbara) rivers together contribute 86% of the waters flowing into the main Nile River (Yacob 2010). In addition, water scarcity (UNDE (2013), stress, and vulnerability are one of Ethiopia's key concerns given its economic development goals and high population growth.

Dams and Barrages on River Channels

Over 45,000 times in the last century, people took the decision to build a dam. Dams were built to provide water for irrigated agriculture, domestic or industrial use, to generate hydropower or help control floods. But dams also altered and diverted river flows, affecting existing rights and access to water, and resulting in significant impacts on livelihoods and the environment. Decisions to build dams are being contested increasingly as human knowledge and experience expand, as we develop new technologies, and as decision-making becomes more open, inclusive and transparent. The World Commission on Dams considers that the end of any dam project must be the sustainable improvement of human welfare. This means a significant advance of human development on a basis that is economically viable, socially equitable, and environmentally sustainable. If a large dam is the best way to achieve this goal, it deserves our support. Where other options offer better solutions we should favor them over large dams. Thus, the debate around dams challenges our view of how we develop and manage our water resources (Dams and development 2000).

Effects of Dams and Barrages on River Channels

Dams and diversions are constructed and operated for a wide variety of purposes including residential, commercial, and agricultural water supply; flood and/or debris control; and hydropower production. Regardless of their purpose, all dams trap sediment to some degree and most alter the flood peaks and seasonal distribution of flows, thereby profoundly changing the character and functioning of rivers. By changing flow regime and sediment load, dams can produce adjustments in alluvial channels, the nature of which depends upon the characteristics of the original and altered flow regimes and sediment loads. Dams disrupt the longitudinal continuity of the river system and interrupt the action of the conveyor belt of sediment transport. Upstream of the dam, all bed load sediment and all or part of the suspended load (depending upon the reservoir capacity relative to inflow) (Brune 1953) is deposited in the quiet water of the reservoir (reducing reservoir capacity) and upstream of the reservoir in reaches influenced by backwater. Downstream, water released from the dam possesses the energy to move sediment, but has little or no sediment load. This clear water released from the dam is often referred to as hungry water, because the excess energy is typically expended on erosion of the channel bed and banks for some years following dam construction, resulting in incision (down cutting of the bed) and coarsening of the bed material until equilibrium is reached and the material cannot be moved by the flows. Reservoirs also may reduce flood peaks downstream, potentially reducing the effects of hungry water, inducing channel shrinking, or allowing fine sediments to accumulate in the bed.

As waters flow from high elevation to sea level, their potential energy is converted to other forms as they sculpt the landscape, developing complex channel networks and a variety of associated habitats. Rivers accomplish their geomorphic work using excess energy above that required to simply move water from one point on the landscape to another. In natural channels, the excess energy of rivers is dissipated in many ways: in turbulence at steps in the river profile, in the frictional resistance of cobbles and boulders, vegetation along the bank, in bends, in irregularities of the channel bed and banks, and in sediment transport. The transport of sand- and gravel-sized sediment is particularly important in determining channel form, and a reduction in the supply of these sediments may induce channel changes. The supply of sand and gravel may be the result of many factors, including changes in land use, vegetation, climate, and tectonic activity. This part of the book is concerned specifically with the response of river channels to a reduction in the supply of these sediments by dams and gravel mining.

The Legal Aspects of the Nile River

What is the legal status of the Nile water treaties described above—or more specifically, is the international legal regime established over the Nile through treaties concluded between Great Britain and Egypt with other powers still operational and binding on Nile Basin states? The answer to this question is fundamental to the issue of rights and obligations over Nile waters.

If the Nile Waters Treaties are valid and binding, they legitimize the legal order of the colonial period that gave Egypt preeminence in the control of the Nile and developments in the basin. This would be a severe constraint on the development efforts and opportunities of upper riparian states. But if the Nile Waters treaties are not binding, then the control and utilization of Nile waters are regulated by the principles of customary international water law. It would also mean that the Nile is in search of a new legal regime in the form of a basin-wide agreement. This would provide plenty of room for negotiation and bargaining as amongst the riparian states. It could help to develop a utilization regime that is more sustainable and equitable (Okoth-Owiro 2004).

Agriculture and Irrigation in Nile Basin

For most of the twentieth century, the Nile River has been the source of political tensions and low-intensity conflicts among three of its major riparian countries (Ethiopia, Sudan and Egypt). However, since the late 1990s, the Nile Basin countries—with the encouragement and support of the international community—have made some attempts to establish basin-wide cooperative institutions. This process of engagement and collaboration is presently under severe stress due to increasing demand and decreasing supply of water resources in the basin. This situation may be complicated further by the global climate change, which is anticipated to result in long-term changes in the volume and pattern of runoff in the Nile River system. Moreover, the emergence of China as a major player in the power politics of the Nile Basin has facilitated a number of unilateral initiatives for large-scale water development projects. In this context, this chapter critically examines the survival and sustainability of water cooperation endeavors in the Nile Basin as the river faces challenges from the global climate change and shifting regional geopolitics (Swain 2011).

Agriculture and Irrigation in Egypt

International Law, in general, exists and effectively contributes to positively settle a wide range of controversial issues. Despite that, it also equally has several more loopholes to strictly examine and mend particularly in major areas of International Water Law. Due to its inherent nature as a rule applying to the most complex interstate relations, international water law obviously could not be as much compressive, to the point and accurately instrumental as required and desired. With particular reference to international rivers, international law undeniably is too general while international rivers are too particularistic. International law tends and seriously tries to be too objective regardless of the unavoidable particularity of each state in question; state interest in international rivers is too subjective. International law tends to be too vague, ambiguous, and idealistic in its provisions and statements, while state use of international rivers demands an approach which is as clearer as possible, straightforward and realistic. The Ethiopian Renaissance Dam represents the fact that state practice in the specific area of international waters is still guided by state practice usually through unilateral actions. As such, international principles, in this case, follow far behind state practice, in the political sense of the term (Habtamu 2011).

Ethiopians now stand as a legal challenger not only to the timely relevance of the traditional Egyptian policy that founded itself on the perceptions of Ethiopia's capacity limitations to make use of the Nile waters but also to the adequacy of international law to preempt interstate misunderstandings. Obviously, the Renaissance Dam has showed that Egyptians have created a formidable duty more on themselves than on Ethiopia by their insistence on pursuing "No Negotiation" Nile Policy. At present, it means that Egyptians, in demonstrating their loyalty to their policy, have to wait patiently until practice proves whether the Dam would actually harm or does not harm their advantages. Legally speaking, Egypt finally finds itself prisoner of its own policy (Habtamu 2011).

Efficiency Use of Irrigation Water

The Nile River in Egypt is the primary source of water. It is used for strategically important water uses such as drinking, fishing, industrial use, livestock, and irrigation. The water in the Nile in Egypt is intricately managed through an extensive system of dams, barrages, and canals. The water from the Nile is conveyed to the users through a vast network of canals. Wastewater and agricultural drainage water from these uses are collected by drains and are often returned to the Nile River as inflows.

Any disruption or impairment to the Nile River from natural or anthropogenic threats can potentially have far-reaching economic and social implications. Approximately 99% of the population of Egypt (total population is approximately

85.783 million) lives within the Nile Valley and Delta, which constitutes less than 4% of Egypt's total area. The area of the Nile Delta itself is about 25,000 km^2 and has approximately 35 million inhabitants (Abdel-Gawad and Khalil 2003).

References

Abrahamsen, R. (2005). Defines securitization following the Copenhagen School as a "political choice, a decision to conceptualize an issue in a particular ways … by an "us against them" distinction and a logic of existential-threat". Rita Abrahamsen, "*Blair's Africa: The politics of securitization and fear", alternatives: Global, local, and political*, p. 60.

Abdel-Gawad, S. and Khalil, B. M., (2003) "Site Registry For The Nile River Water Quality Monitoring Network", Report No.: WQ-TE-0307-011-FN, National Water Quality and Availability Management Project, National Water Quality Monitoring Component.

Habtamu, A. (2011, July). International Legal Perspectives on the Utilization of Trans-Boundary Rivers: The Case of the Ethiopian Renaissance (Nile) Dam. In *Ninth IUCN [International Union for Conservation of Nature] Colloquium, North West University of South Africa, Eastern Cape*.

Arsano, Y. (2010). Institutional development and water management in the Ethiopian Nile Basin. In Tvedt, T. (Ed.), *The River Nile in the post-colonial age* (p. 161). I. B. Tauris & Co. Ltd.

Batstone, R. K. (1959). "The Utilization of the Nile Waters" 8 I.C.L.Q 523

Brune, G. M. (1953). The trap efficiency of reservoirs. *Transactions of the American Geophysical Union, 34,* 407–418.

Buzan, B, et al (1998). Define 'security' "is about survival … the invocation of security has been the key to legitimizing the use of force, but more generally it has opened the way for the state to mobilize …". In B. Buzan, O. Waever, & J. De Wilde (Eds.), *Security*: *A new framework for analysis* (p. 21). London: Lynne Rienner Publishers.

Cascao, A. (2011). Power relations, conflict, and cooperation in the Eastern Nile River Basin. In S. S. Elmusa (Ed.), *The burden of resources—Oil and water in the gulf and the Nile Basin* (p. 28). American University in Cairo Press.

Godana, B. A. (1985). Africa's Shared Water Resources (London, Frances Pinter).

Hamdy, H. A., & Rasheedy, A. A. (2011). The Nile River and Egyptian foreign-policy interests. In K. G. Adar, & N. A. Check (Eds.), *Cooperative diplomacy, regional stability and national interests: The Nile River and Riparian States* (p. 131). Africa Institute of South Africa.

ILA. (2013). The International Law Association as the Helsinki Rules, the UN Watercourses Convention and the Berlin Rules: Perspectives on International Water Law". *Water Resources Development*, The World Bank, Washington DC (2007) (p. 629). http://dx.doi.org/10.1080/07900620701488562. Accessed on June 18, 2013.

Kandie, D. (1999). *Egypt and the hydro-politics of the Blue Nile River* (p. 142). Michigan State University Press. Accessed on June 5, 2013.

Kameri-Mbote, P. (2007). *Navigating peace: Water, conflict, and cooperation: lessons from the Nile River Basin* (p. 1). Woodrow Wilson: International Center for Scholars.

Kasimbazi, E. (1998). The relevance of sub-basin legal and institutional approaches in the Nile basin. *South African Journal of Environmental Law and Policy*, 5(1), 17–34.

Mason, S. A. (2004). *From conflict to cooperation in the Nile Basin* (p. 1), (Ph.D. Thesis). Swiss Federal Institute of Technology, ETH Zurich.

Mumma Martinon, C. A. (2010). Nile Basin initiative: A possibility of turning conflicts into opportunities. In B. Calas, & C. A. Mumma Martinon (Eds.), *Shared waters, shared opportunities*: *Hydro politics in East Africa* (p. 55), (Ph.D. Thesis). French Institute for Research in Africa (IFRA).

Okidi, C. O. (1982). Review of treaties on consumptive utilization of waters of Lake Victoria and Nile drainage system. *Nat. Resources J.*, 22, 161.

Okoth-Owiro, A. (2004). *The Nile treaty, State succession and international treaty commitments: A case study of the Nile water treaties*. Konrad Adenauer Foundation, Law and Policy Research Foundation, Konrad Adenauer Stiftung and Law and Policy Research Foundation 2004.

Qaddumi. (2008). *Practical approaches to transboundary water benefit sharing*, (pp. 1–19). Overseas Development Institute, London, Working Paper (292).

Swain, A. (2011). Challenges for water sharing in the Nile basin: changing geo-politics and changing climate. *Hydrol. Sci. J.*, 56(4), 687–702

Tesfaye, T. (2011). Ethiopia's position on the Nile water agreements. In K. G. Adar, & N. A. Check (Eds.), *Cooperative diplomacy, regional stability and national interests*: *The Nile River and Riparian States* (p. 78). Africa Institute of South Africa.

The Nile River. (2010). Nile Basin Initiative (NBI). http://www.nilebasin.org/newsite/index.php?option=com_content&view=category&layout=blog&id=36&Itemid=75&lang=en. Accessed on June 10, 2013.

UNDE. (2013). Water Scarcity. 2007. http://www.un.org/waterforlifedecade/scarcity.shtml. Accessed June 5, 2013.

WCD, World Commission on Dams. (2000). Thematic review IV. 2 irrigation options. Yields from irrigated areas are on average double that of rain fed agriculture, and are generally higher on land irrigated by ground water than on land irrigated with surface water.

Woldetsadik, T. K. (2013). *International watercourses law in the Nile River Basin*: *Three states at a crossroads* (p. xv). London: Routledge.

Yohannes, O. (2013). Hydro-politics in the Nile Basin: In search of theory beyond realism and neo-liberalism. Journal of Eastern African Studies, 78. http://dx.doi.org/10.1080/17531050802682788. Accessed June 10, 2013.

Chapter 2
The Nile River

Abstract This chapter provides a brief overview of background of the Nile River, scope of the chapter, geographical features of the Nile River, hydrological features of the Nile River, economic aspects of the Nile River, and the legal aspects of the Nile River.

Keywords Scope of the chapter · Geographical features of the Nile River · Hydrological features of the Nile River · Economic aspects of the Nile River · The legal aspects of the Nile River

The Nile River is considered the main life artery for so many African countries especially Egypt. Therefore, it is of the essence to preserve its water and utilize it very efficiently. The hydrological system of the Nile River under consideration has several dams and barrages that are equipped with control gates. The improvement of these hydraulic structures' criteria for operation can be assessed if reliable forecasts of inflows to the reservoir are available.

Background

The Nile, which is the only drainage outlet from Lake Victoria, is one of the longest rivers in the world. Its total length together with those of its tributaries is about 3,030,300 km. The catchment area of the Nile totals some 2,900,000 km^2, representing about one-tenth of the surface area of the entire African Continent (Okoth-Owiro 2004).

The Nile measures some 5611 km from its White Nile source in Lake Victoria (East Africa) and some 4588 km form the Blue Nile source in Lake Tsana (Ethiopia). Thus, the river system originates from two distinct geographical zones (Okoth-Owiro 2004).

One subsystem, with the White Nile as its main artery, originates in the equatorial lakes of East and Central Africa, the most important of which is Lake Victoria, and in the Bahr-el-Ghazal water system—a vast lagoon formed by the

© Springer International Publishing AG 2017

Y.M. Hamada, *The Grand Ethiopian Renaissance Dam, its Impact on Egyptian Agriculture and the Potential for Alleviating Water Scarcity*,
Environment & Policy 55, DOI 10.1007/978-3-319-54439-7_2

convergence of a number of streams rising to the East and North of the Nile-Congo divide (Okoth-Owiro 2004).

The other subsystem consists of the Blue Nile and its tributaries, the Atbara and the Sobat. It originates from the Ethiopian Plateau (Okoth-Owiro 2004).

The Nile is made up of three main tributaries. These are the White Nile, the Blue Nile, and the Atbara. The White Nile rises from its source in the highlands of Rwanda and Burundi and flows into Lake Victoria. It leaves Lake Victoria at its northern shore near the Ugandan town of Jinja, through a swampy stretch around Lake Kyoga in Central Uganda and then heads north toward Lake Albert (Okoth-Owiro 2004).

Lake Albert receives a good amount of water from the Semliki River, which has its source in the Congo and empties first into Lake Edward, where it receives additional water from tributaries coming from the Ruwenzori Mountains on its way to Lake Albert. From Lake Albert, the White Nile flows north into Southern Sudan (Kasimbaji 1998).

Lake Victoria, Edward, and Albert are the natural reservoirs, which collect and store great quantities of water from the high rainfall regions of Eastern Equatorial Africa and maintain a permanent flow down the White Nile with relatively small seasonal fluctuations (Okoth-Owiro 2004).

In Southern Sudan, near the capital city of Khartoum, the White Nile meets the Blue Nile, which drains Lake Tsana in the Ethiopian Highlands. The two flow together to just north of Khartoum where some 108 km downstream, and they are joined by the Atbara, the last important river in the Nile system, whose source is in Eritrea. The river then flows north through Lake Nasser and the Aswan Dam before splitting into major distributaries, the Rosetta and Damietta, just north of Cairo. These distributaries flow into the Mediterranean Sea (Okidi 1982).

According to Godana (1985), the average annual flow of the Nile is 84 milliards of cubic meters as measured in Aswan. Of this total, Batstone (1959) estimates that 84% is contributed by Ethiopia and only 16% comes from the Lake Plateau of Central Africa. A similar distribution pattern is given by Godana (1985) who asserts that 85% of the flow of the Nile originates from the Ethiopian plateau, whereas only 15 only comes from the East African source areas. It is, however, important to note that the statistics of the flow of the Nile are a complex matter, which the above estimates tend to over simplify.

Scope of the Chapter

This chapter focuses on the existing debates of "the Nile River." In general, the waters of the Nile are utilized for irrigation, hydroelectric power production, water supply, fishing, tourism, flood control, water transportation, and the protection of public health (Kasimbazi 1998). In particular, it should be noted that the economy of the entire Nile Basin almost entirely consists in the agricultural activities of the co-riparian of the Nile—Rwanda, Burundi, Congo, Tanganyika, Kenya, Uganda,

South Sudan, Sudan, Ethiopia, Eritrea, and Egypt. In the upper-basin states of Ethiopia, Eritrea, Kenya, Uganda, Tanzania, Democratic Republic of Congo, Rwanda, and Burundi, settled agriculture is the general economic activity. The lower-basin states of Sudan and Egypt are also primarily agricultural economies but, in contrast to the upper-basin states, their agriculture is largely irrigation based. The economic use of the Nile for the purposes of agriculture (particularly irrigation-based agriculture) is, therefore, its most important use (Okoth-Owiro 2004).

In Egypt, a desert agricultural country, the entire life of the nation is dependent on the river's waters. As President **Anwar Sadat** stated in 1978, "**We depend upon the Nile 100% in our life, so if anyone, at any moment, thinks of depriving us of our life we shall never hesitate to go to war**" (Kukk and Deese 1996).

The Nile is one of the world longest rivers, flowing from south to north with a length of 6850 km, over 35° of latitude. Its catchments basin covers approximately 10% of the African continent, with an area of 2.9 Million km^2 (Godana 1985; Said 1993). The Nile Basin extends from latitude of 4° South to 31° North and includes 11 different countries: Burundi, Egypt, Eritrea, Ethiopia, Kenya, Rwanda, Sudan, South Sudan, Tanzania, Uganda, and the Democratic Republic of Congo.

The basin of the Nile is characterized by the existence of two mountainous plateaus rising some thousands of meters above mean sea level. The Equatorial or Lake Plateau in the southern part of the Nile basin, situated between the two branches of the Great Rift, is at a level of 1000–2000 m and has peaks of 5100 and 4300 m. This plateau contains Lakes Victoria, George, Edward (Mobutu Sese Seko), and Albert, which slope gently toward the north at an average rate of one meter for every 20–50 km of stretch. In contrast, the rivers which connect these lakes fall at an average rate of one meter every kilometer or less of length. The Ethiopian or Abyssinian Plateau, which forms the eastern part of the basin, has peaks rising to 3500 m. North of the Lake plateau the basin descends gradually to the Sudan plains where the Nile runs at altitudes lower than 500 m in its northerly direction (Godana 1985; Said 1993).

The Nile enters the Mediterranean Sea through a delta that separates the river into the Rosetta and Damietta distributaries. Not only does the Nile provide freshwater to millions, but within its basin there are five major lakes with a surface area totaling more than 1000 km^2 (Victoria, Edward, Albert, Kyoga, and Tana), vast areas of permanent wetlands and seasonal flooding (the Sudd, Bahr al-Ghazal, and Machar marshes), five major reservoir dams (Aswan High Dam, Rosaries, Khashm El Girba, Sennar, and Jebel Aulia), and three hydroelectric power dams (Tis Isat, Finchaa, and Owen Falls). The entire Nile Basin consists of eight major sub-basins with very different physical, hydrologic, and climatic characteristics (Abdel Aty et al. 1999).

The Nile River is the longest international waterway in the world. Its tributaries flow across 11 countries in northeastern Africa before reaching the Mediterranean Sea. Lake Victoria and other tributaries whose waters originate in Rwanda, Burundi, Congo, Tanzania, Kenya, and Uganda feed into the White Nile, which flows 5584 km northward through Uganda, South Sudan, and into North Sudan where it meets the Blue Nile at Khartoum. Originating in Lake Tana, situated six thousand feet above the sea level in the Ethiopian highlands, the Blue Nile gathers its waters from tributaries in Eritrea, Ethiopia, and Sudan, and covers a distance of 1529 km. From the confluence of the White and Blue Nile at Khartoum, the Nile flows northward into Egypt and on into the Mediterranean Sea. At 84 billion m^3, the Nile's average annual runoff is modest in comparison to the other major river systems in Africa, while the demand for its water is much higher. The figure for the Congo River is 1250 billion m^3; for the Niger River, 180 billion m^3; for the Zambezi River, 230 billion m^3; and for the Volta River, 390 billion m^3 (Rangeley 1994).

The White Nile provides a small but steady flow of water, fed by the snows of the Ruwenzori Mountains in Western Uganda as well as Lake Victoria and other, smaller equatorial lakes. The Equatorial Lakes Region, however, accounts for only 14% of the Nile's yearly flow. This is partly because the White Nile loses a considerable amount of water to swamp areas near its source, and then to evaporation as it proceeds northward into the arid regions of Sudan. The Blue Nile, on the other hand, contributes 86% of the yearly flow, but is characterized by significant seasonal fluctuations. The Blue Nile itself accounts for 59%, while the Baro-Akobo (Sobat) provides 14%, and the Tekezze (Atbarah) 13% of the yearly flow, (Swain 1997). Indeed, during the flood season, up to 95% of the Nile runoff originates in the Ethiopian highlands, which only serves to increase Egypt's dependence on its neighbor for freshwater resources during that time of the year. Moreover, the average annual flow of the Nile has declined over the years. At Aswan, where Egypt constructed a gargantuan dam in the 1960s, the average annual flow was estimated at 110 billion m^3 in the late nineteenth century. It has declined significantly to just above 80 billion m^3 since then, though Egypt's demand for fresh water in recent years has increased dramatically (Ezzat et al. 1996).

Egypt is totally dependent on water from the Nile River. Table 2.1 demonstrates irrigation and hydropower potential of Nile basin countries. Egypt has already been able to exploit most of the irrigation and hydropower potential available within its

Table 2.1 Irrigation and hydropower potential of Nile basin countries

Country	Area (1000 km^2)	Arable land (1000 ha)	Irrigated area (1000 ha)	Irrigation potential (1000 ha)	Hydropower potential (MW)	Hydropower installed (MW)
Burundi	26	770	14	185	1366	36
Congo	2345	6930	11	n.a.	530,000	2829
Egypt	1001	2800	3266	4434	3210	2825
Eritrea	118	n.a.	28	n.a.	n.a.	n.a.
Ethiopia	1104	11,300	190	3637	162,000	378
Kenya	583	4000	67	352	30,000	611
Rwanda	26	8504	4	160	3000	59
Sudan	2506	12,920	1946	4843	1900	225
Tanzania	945	3100	190	828	20,000	339

Source African Development Bank, Policy for Integrated Water Resources Management, February 2000

own borders. Riparian countries such Ethiopia and Sudan are very much behind in this race. Economic and political difficulties in Ethiopia and Sudan have so far hampered the development of water resources, such as the construction of dams and other water infrastructure projects. Both countries, and especially Ethiopia, which is seen as the "water tower" of the basin, face an increasing demand for fresh water resources and are therefore exploring different options for developing the Nile tributaries flowing through their territories.

The first agreement exclusively dealing with sharing and allocating the water of the Nile was signed in 1929 between Sudan, represented by Great Britain at the time, and Egypt. The agreement allocated 48 billion m^3 of water to Egypt and 4 billion to Sudan. From the early 1930s, Sudan gradually adopted irrigational agriculture and the demand for water increased (Wallenstein and Swain 1997). As a result, on the eve of its independence and following the Egyptian revolution in 1952, the administration in Sudan started demanding that the agreement with Egypt be renegotiated. After a period of bilateral tension, negotiations resumed and a new agreement was signed in 1959 (Warburg 1991). Based on new calculations that put the annual runoff at 82 billion m^3, Egypt received 55.5 billion m^3, 18.5 billion being allotted to Sudan (the rest being lost to evaporation and swamp areas). The new agreement also included some provisions regarding the replenishment of the reservoir of the planned Aswan Dam (Christiansen 1992).

The construction of the Aswan Dam began in 1960, immediately after the two countries signed the agreement, and was completed in 1971. For the next 25 years, until the fall of Sudanese President Numayri in 1985, Egypt had a friendly regime in Sudan. In return for helping Numayri stay in power, Sudan agreed to construct the Jonglei Canal, which was scheduled to begin in 1978. As a result of population growth and continuing drought throughout the upper basin, the Egyptian government was keen to enhance the water supply of Lake Nasser, which had experienced a noticeable decrease in the late 1970s and early 1980s (Warburg 1985). The canal was specifically designed to decrease the loss of water from the White Nile when it traversed the Sudanese swamps (Collins 1988). The Sudanese wetlands in Southern Sudan are one of the largest swamps in the world. The Jonglei canal system would have diverted annually up to 4.7 billion m^3 of water from the wetlands back to the Nile, of which 3.8 billion were allocated to Egypt for augmenting the water supply of Lake Nasser. However, opposition from the Sudanese People's Liberation Army brought the digging of the canal to a halt in 1984. It was widely believed in the south of Sudan that the Jonglei project was being implemented to benefit exclusively the northern part of the country as well as Egypt at the expense of the south (Swain 2011).

Geographical Features of the Nile River

The Nile basin, because of its size and variety of climates and topographies, constitutes one of the most complexes of all major river basins Koran and Barrett (1994). Figure 2.1 shows a map of the River Nile Basin (World Bank Publication 2008).

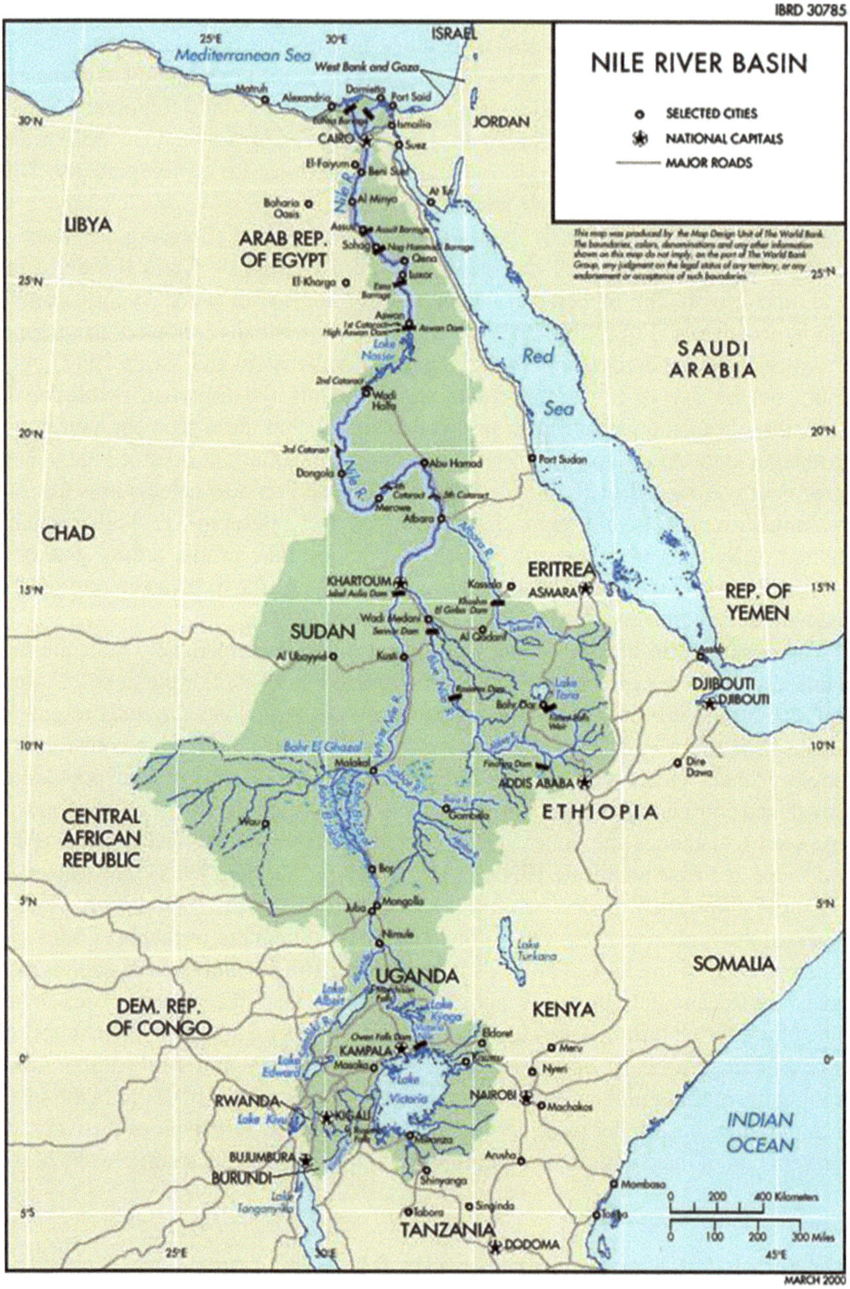

Fig. 2.1 Map of the River Nile basin. *Source* World Bank Publication (2008)

Tributaries: The Nile is made up of three main tributaries. These are the White Nile, the Blue Nile, and the Atbara.
Major Regions: Researchers believe that the Nile originated 30 million years ago in the mid-Tertiary period. Its headstream was probably the Atbara River. The river basin continued to evolve and now has seven major regions:

1. Lake Plateau of East Africa
2. Al-Jabal (Mountain Nile)
3. White Nile
4. Blue Nile
5. Atbara
6. United Nile (North of Khartoum in the Sudan and Egypt)
7. Nile Delta.

Sources: The White Nile originates from the Lake Plateau region of East Africa where several headstreams contribute to the Lake Victoria and Lake Albert. The Ruvyironza, regarded as the ultimate source of the Nile, is one of the upper branches of the Kagera River. The Blue Nile gathers its volume mainly from the Ethiopian upstream of Lake Tana, some 2150 m (or 7054 ft.) above sea level.
Length: The River Nile is the longest river in the world. From Lake Victoria to the Mediterranean Sea the length of the Nile is 5584 km (3470 miles). From its remotest headstream, the Ruvyironza River in Burundi, the river is 6671 km (4145 miles) long. From Lake Tana in Ethiopia to the Mediterranean Sea, the Nile is 4588 km long. The total length of the Nile, together with its tributaries, is about 3,030,300 km.
Basin Area: There are 11 basin countries on the Nile. These are Egypt, Sudan, South Sudan, Ethiopia, Eritrea, Uganda, Tanzania, Kenya, the Democratic Republic of Congo, Rwanda, and Burundi. The area of the Nile Basin is 3,030,700 km^2, distributed as follows:

1. Sudan and South Sudan—1,900,000 km^2 (62.7%)
2. Ethiopia and Eritrea—368,000 km^2 (12.1%)
3. Egypt—300,000 km^2 (9.9%)
4. Uganda—232,700 km^2 (7.7)
5. Tanzania—116,000 km^2 (3.8%)
6. Kenya—55,000 km^2 (1.8%)
7. D.R. Congo—23,000 km^2 (0.8%)
8. Rwanda—21,500 km^2 (0.7%)
9. Burundi—14,500 km^2 (0.5%).

The population of the basin area is estimated to be about 160 million, while the total population of all the basin countries is about 300 million.
Dams: The major dams on the Nile are Roseires Dam, Sennar Dam, Aswan High Dam, and Owen Falls Dam.

Hydrological Features of the Nile River

The average discharge of the Nile is about 300 million m^3 per day. Godana (1985) reports that as measured at Aswan the average annual flow of the Nile is 84 milliards of cubic meters. Of this total, Batstone (1959) estimates that 84% is contributed by Ethiopia and only 16% is came from the Lake Plateau of East and Central Africa. Garretson (1960) and Godana (1985) provide similar estimates. As the flow chart diagram, while the flow of the White Nile is relatively regular and stable throughout the year, the flow of both Blue Nile and the Atbara subsystems fluctuates seasonally.

(a) **Flow rates**:

At its peak discharge in August to October, the Blue Nile swells to an enormous torrential flow and accounts for some 90% of the waters passing Khartoum. But by April, it will have dwindled to one fortieth of the flood discharge, to account for no more than 20% of the waters passing Khartoum.

Garretson (1960) estimates that at the peak of its flood, the Blue Nile alone supplies 90% of the water passing through Khartoum, but in the low season, it provides only 20%. Thus, according to Godana (1985), the White Nile at Khartoum provides only 40% of the river's peak discharge, but at the low flow accounts for four fifths of the total delta discharge.

It has also been estimated that on average, 59% of the Nile flow is from the Blue Nile, 28% from the White Nile, and 13% from the Atbara River (Carroll 1999).

Fourteen percent of the White Nile flow is from the upper Nile states and the other 14% is from the Sobat River. The Blue Nile's contribution is the largest but it is seasonal. Most of the flow comes in August, September, and October just after the monsoon season in the Ethiopian highlands. At those times, the Blue Nile may account for up to 90% of the Nile flow, whereas in July, just prior to the wet season, it may account for as little as 20% of the main flow of the Nile (Okoth-Owiro 2004). The White Nile's contribution,→ on the other hand, is small but steady.

The contribution to Nile flow also varies considerably among Countries. Ethiopia contributes 86% of Nile flow, whereas Egypt contributes nothing. The annual discharge of the Nile in units of 12 milliards of cubic meters has been calculated by Garretson (1960) as follows:

White Nile Downstream at Lake Albert in Uganda	2
White Nile Downstream of the Sudd in Southern Sudan	1
Sobat from Southwest Ethiopia	1
White Nile at Khartoum	2
Blue Nile at Khartoum	4
Atbara from Northwest Ethiopia	1
Main Nile at Sudanese/Egyptian Border	7

(b) **Loss of water**:

Significant amounts of water are lost in the Nile basin through evaporation and soakage. Some 812 billion cubic feet of water is brought into Lake Victoria by those rivers that drain into it. This represents 15% of the water entering the lake, the other 85% doing so from precipitation directly onto the lake surface. Evaporation helps to balance the water that drains into the lake and continues to Lake Albert. Some 85% of the water leaving Lake Victoria does so through direct evaporation from surface and only the remaining 15% leaves by way of the Victoria Nile, which leaves the lake near Jinja in Uganda, and flows via the Owen Falls, Lake Kyoga, and the Murchison Falls to join the outflow from Lake Albert (Okoth-Owiro 2004).

Godana (1985) estimates that some 24 milliards cubic meters of water flow down the White Nile from Lake Albert and the East African highlands, half of which is lost through intense evaporation and soakage in the Sudan.

An official Sudanese Government study puts the total swamp losses at 42 milliards of cubic meters (Sudan 1975). In fact, the Jonglei Canal project was intended to divert the flow of the Nile in Southern Sudan to avoid the enormous evaporation losses, which occur there. Lake Nasser, the second largest manmade lake in the world and the result of the Aswan Dam project, loses 10% of its volume annually through evaporation. This is because of its location in the middle of the desert (Okoth-Owiro 2004).

Economic Aspects of the Nile River

Practically all the Nile basin states view the Nile as a principal feature of their economies. The Nile and its resources are used for irrigation (principally by Egypt and the Sudan), hydroelectric power generation, water supply, fishing, tourism, water transportation, and the protection of public health (Okoth-Owiro 2004).

Egypt is the most dependent country on the Nile. The Nile provides Egypt with an average of 55.5 billion m^3 of water, or 86% of the country's usable water. The Aswan Dam alone produces 1/3 of Egypt's electricity. It is estimated that Egypt relies on the Nile for 95% of its water needs (Okoth-Owiro 2004).

Ethiopia is planning to construct, or is in the process of constructing, a new facility on the Blue Nile in order to supply irrigation water for 1.5 million settlers in the Western province of Welega and to provide a steady source of hydroelectric power for the country. The facility is expected to divert 39% of the Blue Nile's waters (Okoth-Owiro 2004).

Sudan has also been increasing the number of projects it is undertaking on the Nile with the use of the Nile waters. These include irrigation, dam construction, hydroelectric power generation and canalization. It is the second most dependent country on the waters of the Nile. There is general evidence of increased utilization of Nile waters among the riparian states. This is raising the prospects of conflict and water-scarcity in the region (Okoth-Owiro 2004).

The complete control of the river over the economy of Egypt has been characterized as the unique feature of the Nile, setting it apart from all other international rivers (Pompe 1958).

Increasingly, all the basin states have come to view the Nile as a principal feature of their economies. They show an increasing interest in the abstraction and diversion of Nile water for various development purposes, irrigation included. Examples of such developments include the Jonglei Canal Project in Sudan (which has been dormant due to the raging conflict since 1983), and the planned construction, by Ethiopia, of a new facility on the Blue Nile to supply irrigation water for 1.5 million newly resettled peasants in the western province of Welega as well as to provide a steady source of hydroelectric power for the country (Kukk and Deese 1996).

Godana (1985) also reports that Tanzania hopes to implement a plan to abstract the waters of Lake Victoria to irrigate the relatively low and dry steppes of Central Tanzania. And with the establishment of the Lake Basin Development Authority in Kenya, the country has begun to treat the resources of the Lake Victoria basin more comprehensively.

The economic importance of the Nile is also reflected in the establishment of various sub-basin initiatives for the development and management of basin resources. These include the Kagera Basin Organization, the Technical Cooperation for the Promotion of the Development and Environmental Protection of the Nile Basin (TECCONILE), and the Lake Victoria Environment Management Programmed (LVEMP) (Kasimbazi 1998).

The Legal Aspects of the Nile River

The Nile is an international river. As a shared water resource, the development, utilization, and management of the Nile basin waters are regulated by international water resources law. Following the nomenclature of Article 38 of the Statute of the International Court of Justice, international water resources law may be derived from:

(a) International conventions, whether general or particular;
(b) International customs;
(c) The general principles of law recognized by civilized nations; and
(d) As a subsidiary means, the judicial decisions and the teachings of the most highly qualified publicists of the various nations.

The import of the conclusion that the Nile Water Treaties are no longer binding and operational is that only post-colonial agreements can be said to be valid. The only such treaty is the 1959 agreement for the full utilization of the Nile waters. This agreement is a bilateral arrangement between Sudan and Egypt, which does not bind or affect the other riparian states of the Nile. Thus, the legal regime governing the utilization of the waters of River Nile is customary international law. As the international law commission advised: "In the absence of bilateral or multilateral agreements, member states should continue to apply generally accepted principles of international law in the use, development and management of shared water resources" (Biswas 1993: 175). What then does international law say about shared water resources?

References

Abdel Aty, M., Anter, M., & Attia, B. (1999). Nile forecast system: Methodology, database and outputs. In *Proceedings Seventh Nile 2002 Conference, Comprehensive Water Resources Development of the Nile Basin: The Vision for the Next Century*. Cairo, Egypt.

Batstone, R. K. (1959). The utilization of the Nile waters. *International and Comparative Law Quarterly, 8,* 523.

Biswas, K. (1993). *Development of an axisymmetric phase change finite element model for evaluation of roof support systems in arctic placer mines* (M.S. thesis, University of Alaska Fairbanks).

Carrol, C. M. (1999). Past and future legal framework of the Nile River Basin. *Georgetown International Environmental Law, 12,* 269.

Christiansen, S. (1992). Shared benefits, shared problems. In S. Lodgaard & A. H. af Ornäs (Eds.), *The environment and international security* (p. 57). Oslo: PRIO Report No. 3.

Collins, R. O. (1988). *The waters of the Nile: Hydro politics of the Jonglei Canal, 1900–1988*. Oxford: Clarendon Press.

Ezzat, M. N., Mohamadien, M. A., & Attia, B. B. (1996). Integrated approach to water resources development. In *Arab Republic of Egypt's Country Paper Presented at the Nile 2002 Conference*. Kampala, Uganda.

Garretson, A. H. (1960). The Nile River System. In *Proceedings of the American Society of International Law*. Washington, DC.

Godana, B. A. (1985). *Africa's shared water resources*. London: Frances Pinter and Boulder, Colorado: Lynne Rienner Publishers.

Kasimbazi, E. (1998). The relevance of sub-basin legal and institutional approaches in the Nile basin. *South African Journal of Environmental Law and Policy, 5*(1), 17–34.

Koran, V., & Barrett, C. B. (1994). NOAA/NWS/OH/MFS project silver spring, Md, satellite based distributed monitoring, forecasting, and simulation (MFS) system for the Nile River application of remote sensing in hydrology. In *Proceedings of the Second International Workshop, NHRI Symposium No. 14*. Saskatoon: NHRI.

Kukk, C.L., & Deese, D. A. (1996). At the water's edge: Regional conflict and cooperation over fresh water. *UCLA Journal of International Law and Foreign Affairs, 1,* 21–64.

Okidi, C. O. (1982). Review of treaties on consumptive utilization of waters of Lake Victoria and Nile drainage system. *Natural Resources Journal, 22,* 161.

Okoth-Owiro, A. (2004). The Nile treaty, state succession and international treaty commitments: A case study of the Nile water treaties. Konrad Adenauer Foundation, Law and Policy Research Foundation, Konrad Adenauer Stiftung and Law and Policy Research Foundation.

Pompe, C. A. (1958). The Nile Waters Question. Symbolae Verzijl, the Prague.

Rangeley, R., Thiam, B. M., Andersen, R. A., & Lyle, C. A. (1994). *International river basin organizations in Sub-Saharan Africa*. Washington, DC: World Bank Technical Paper No. 250.

Said, R. (1993). *The River Nile, geology, hydrology and utilization*. Pergamum: Former head of the geological survey of Egypt.

Sudan, Rep. Of. (1975). *Control and use of the Nile water in the Sudan*. Sudan: Ministry of Irrigation, Khartoum.

Swain, A. (2011). Challenges for water sharing in the Nile basin: changing geo-politics and changing climate. *Hydrological Sciences Journal–Journal des Sciences Hydrologiques*, *56*(4), ISSN: 0262-6667 (Print) 2150-3435 (Online) Journal homepage: http://www.tandfonline.com/loi/thsj20..

Swain, A. (1997). The Nile River dispute: Ethiopia, the Sudan, and Egypt. *Journal of Modern African Studies, 35*(4), 675–694.

Wallenstein, P., & Swain, A. (1997). *Comprehensive assessment of the freshwater resources of the world, International fresh water resources: Conflict or cooperation?* Stockholm: Stockholm Environment Institute.

Warburg, G. R. (1985). *Egypt and the Sudan: Studies in history and politics* (p. 566). London: Frank Cass.

Warburg, G. R. (1991). The Nile in Egyptian-Sudanese relations. *Orient, 32*(4), 570.

World Bank Publication. (2008). *Africa–Nile Basin initiative institutional strengthening project: Environmental assessment* (Vol. 1 of 3). Strategy for addressing environmental and social safeguards, report No. E1878.

Chapter 3
The Nile River Basin and Water Security

Abstract This chapter provides a brief overview background of the Nile River Basin and water security, scope of the chapter, geographical snapshot of the Blue Nile River, the Nile River Basin, puzzles and theoretical understandings, Ethiopian hegemony, and hydro-hegemony (Egypt and Sudan).

Keywords Scope of the chapter · Geographical snapshot of the Blue Nile River · The Nile River Basin · Puzzles and theoretical understandings · Ethiopian hegemony and hydro hegemony (Egypt and Sudan)

The main differences between Egypt's hegemonic water management policy and Ethiopia's more cooperative policy focused on poverty reduction can be summarized as follows [since poverty and a lack of water are linked (Water Project 2013)]:

- Ethiopia has enacted a national water resources management policy in 1999 for the first time in its history, following a modern approach to water usage based on the "equitable", "sustainable", and "efficient" measures (Arsano 2010). Egypt's National water policy of 1975 has not yet been updated according to the current hydro-political and environmental shift, which seeks to demand a cooperative action (NBI 2010). This is against the milieu of many experts' call for immediate action to the profound social consequence of freshwater scarcity and violence (Homer-Dixon 1999).
- Echoing the new contemporary national policy, Ethiopia is challenging Egypt's hydro-hegemony via the Nile Basin Initiative (NBI 2010) framework as a basis for creating mutually acceptable outcomes.

Background

Water has always been a highly contentious issue in regional and international affairs, because of its importance for political and socioeconomic stability of nations. Without access to water, the maximization and protection of human and

© Springer International Publishing AG 2017
Y.M. Hamada, *The Grand Ethiopian Renaissance Dam, its Impact on Egyptian Agriculture and the Potential for Alleviating Water Scarcity*,
Environment & Policy 55, DOI 10.1007/978-3-319-54439-7_3

economic development is difficult, if not impossible. The management of this scarce resource in equitable and sustainable terms is currently an important debate within the Nile Basin-wide region, addressing the issues of poverty, population growth, soil erosion, environmental degradation, and climate change. The Nile River waters are highly important to the evolving conflict between eleven riparian states: Kenya, Sudan, South Sudan, Burundi, Eritrea (observer status), Republic of Congo, Rwanda, Uganda, Tanzania, Ethiopia, and Egypt. Figure 3.1 shows the Nile River Basin watershed and its 11 countries at a crossroads.

Egypt and Sudan, perceived as hydro-hegemons, continue to exhibit a growing concern over Ethiopia's move to divert some of the waters of the Blue Nile River for the purpose of hydroelectricity. This means that both countries' favorable colonial-legal agreements, signed between Egypt, Sudan, and Britain in the allocation of the Nile Waters, could be in jeopardy (Cascao 2011). Especially, Egypt's

Fig. 3.1 The Nile River Basin watershed and its 11 countries at a crossroads. *Source* Stratfor (2013)

principal life sustenance depends on the Nile River (Hassan and Rasheedy 2011), and one of Egypt's geographical challenges is the inhabitable desert and nearly zero rainfall. **Daniel Kandie** stresses that, the Nile then drains into Egypt—a country where there is practically no rain, and 86% of the land is classified as very arid, and the rest as arid (Kandie 1999). Water, for Egypt and for other riparian states, is thus often part of discourses and practices of "securitization" (Abrahamsen 2005) and "militarization" guided by fear, mistrust and "zero-sum" politics. Paradoxically, what is an issue of national "security" (Buzan et al. 1998) and life and death for Egypt, is also so for all other riparian states. **Okbazghi Yohannes** places the recently growing concerns into perspective by arguing that now all Nile countries define water resources in terms of national security, hence the elevation of water resources to the status of high politics (Yohannes 2013).

As recently as June 2013, several Egyptian politicians expressed the familiar political rhetoric of a "zero-sum" game to destabilize Ethiopia and the region through various political-military-intelligence maneuvers. Others, including the President of Egypt Mohamed Morsi and Freedom and Justice Party Chairman Saad Al-Katatni, vowed to first-use diplomatic avenues and second rely on the international law to protect Egypt's water security (Egypt President politicians plotting against Ethiopia's dam English subtitle 2013). All the bellicose speeches were directed at the Grand Ethiopian Renaissance Dam (**GERD**) construction over the Blue Nile River in the northwest of Ethiopia.

Scope of the Chapter

This chapter focuses on the existing debates of "water security in the Nile River Basin" and the paradigm shifts of long-term sustainable, manageable, and equitable water-sharing arrangements within Ethiopia and within all the riparian states. Ideally, this would mean the emergence of regional economic cooperation and integration. I have chosen this topic to draw attention to the fundamental changes taking place within Africa, and more specifically within Ethiopia. Moreover, the topic could not be more timely appropriate, since the issue (as of 2013) has reached a critical moment of trans-boundary relations on access to the Nile Waters (Woldetsadik 2013). Some of the key changes seen in the last decade are the constant increase of population and economic growth in all riparian states, particularly in Ethiopia.

In recent years, Ethiopia has come out strong against the asymmetric power relations governing access and utilization of the Blue Nile waters within its own territory. Ethiopia's unilateral action for hydraulic development on the Blue Nile River is supported by the Berlin, Helsinki and International Law Commission (ILA 2013) rules on "equitable" and "no-harm" water usage (Tafesse 2011). The Nile waters, thus, are not only about "rights" to access and utilization, but also about managing the scarce resource equitably and sustainably with all riparian states involved.

Ethiopia is building the momentum to create a sense of urgency of the issue of unequal water usage distribution and management against Egypt and Sudan, using unilateral, bilateral, and multilateral strategies. As Egypt continues to stress its dependence on the Nile, it is considered a matter of national security for Egypt not to accept a genuine basin-wide cooperative legal framework (Mumma Martinon 2010). Contrary to this old and static view, Ethiopia has proceeded unilaterally to utilize some of the Blue Nile waters for domestic and regional-wide electricity production. One of the many reasons for Ethiopia's initial move is that many of Ethiopian rivers, the Abbay (Blue Nile), Baro-Akobo (Sobat), and Tekeze (Atbara) rivers, together contribute 86% of the waters flowing into the main Nile River (Arsano 2010). In addition, water scarcity (UNDE 2013), stress, and vulnerability are one of Ethiopia's key concerns given its economic development goals and high population growth. Figure 3.2 shows water scarcities in Africa.

However, this new political and legal development has not been welcomed by Egypt and Sudan. They both greatly benefit from the 1959 bilateral legal agreements that give them 55.5 bcm (billion cubic meters) of water for Egypt and 18.5 bcm for Sudan, while Ethiopia receives zero (Cascao 2011). **Ana Cascao** places the latter in perspectives by questioning further the old way of doing business: Does the legal and technical control of the downstream countries over the Nile waters that both Egypt and Sudan exerting also mean consistent political control over the waters? (Cascao 2011). Keeping such an important question in mind, the main argument of this point is that Ethiopia, as the primary shareholder of the Nile River Basin, continues to influence and challenge the hydro-hegemonic order of Egypt and Sudan through various counter-hegemonic strategies for domestic and regional-wide economic development interest, cooperation, and integration.

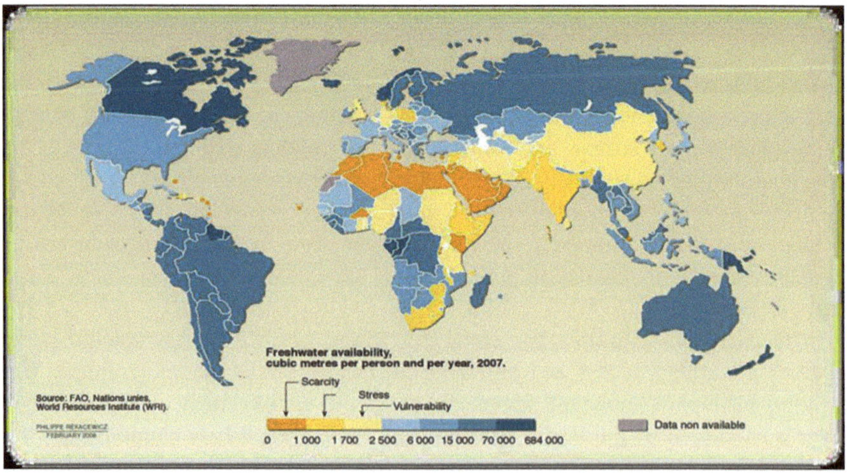

Fig. 3.2 Water scarcities in Africa. *Source* UNDE Water Scarcity (2007)

Geographical Snapshot of the Blue Nile River

The Blue Nile (in Amharic the Abbay River) originates in Ethiopia's northwestern plateau (Arsano 2013). Its waters come from various small- and medium-sized rivers within Ethiopia including Lake Tana and the rivers Dabus, Didessa, Fincha, Guder, Muger, Jamma, Wolaka, Bashilo, Birr, Beles, Dinder, and Rahad. Lake Tana, with its heart-shaped form Fig. 3.3 shows the Blue Nile (Tis Issat) falls, Ethiopia, (Arsano 2010) is the main tributary of the Blue Nile River that flows into Sudan and then into Egypt. The Blue Nile provides a vital source of freshwaters to Sudan and Egypt (Conway 2009). It raises 1850 above sea level and proceeds roughly for 992 km to the Ethio-Sudanese Border (Woldegebriel 2013) before meeting the White Nile in Khartoum, Sudan. Egypt's annual water consumption depends on 86% of waters the Blue Nile produces, the rest coming from the White Nile (Kandie 2013). Consequently, the Blue Nile River is highly important to Egypt, Sudan, and Ethiopia but also faces major contemporary issues of climate change, environmental degradation, and high variability of rainfall.

Fig. 3.3 The Blue Nile (Tis Issat) Falls, Ethiopia. *Source* NBI (2012)

The Nile River Basin

The River Nile is one of the longest rivers in the world (Mumma Martinon 2010). It flows for 6600 km, traversing more than 30 degrees of latitude (Mason 2004). According to **Simon Mason**, The Nile Basin covers an area of 3.1 million km^2, of which 1% is urban, 2% are covered by forest, 3% by wetlands, 3% by open water, 4% by shrub, 5% by irrigated cropland, 10% by cropland, 30% by desert/semi-desert, and 42% by grassland (Mason 2004). Table 3.1 shows statistical facts on the Nile Basin. Within the Nile Basin area approximately 160 million people depend on the river for their livelihoods (Kameri-Mbote 2007). It also provides life sustenance to human needs and to socioeconomic activities for almost 500 million peoples (as of 2012 census) across the 11 riparian states (Nile River 2013). It flows from the south (upstream states) to the north (downstream states) for 6825 km (Mumma Martinon 2010). Within all the riparian states, the Nile River faces present-day threats such as constant droughts, water pollution from land-use activities (Kameri-Mbote 2007), and population growth. Figure 3.4 shows the Nile Basin. This means that the Nile waters are "international" and they require the immediate attention of sustainable and manageable multilateral water-sharing arrangements.

Table 3.1 Statistical Facts on the Nile Basin

Basin Area	3173×103 km^2
Location	−4°S to 31°N and 24°E to 40°E
Riparian States	Burundi, DR Congo, Egypt, Eritrea, Ethiopia, Kenya, Rwanda, Tanzania, South Sudan, Sudan and Uganda
Main Tributaries	Victoria Nile/Albert Nile, Bahr El Jabel, White Nile, Baro Pibor-Sobat, Blue Nile, Atbara, Bahr El Ghazal
River Length	6850 km
Estimated Navigable Length	4149 km
Major Lakes with in the Basin	Lake Victoria, Lake Tana, Lake Kyoga, Lake Albert
Population (Total in all the Nile Countries)*	437 Million
% Population within the Nile Basin*	54% (238 Million)
Temperature	Night Minimum −10 °C and daily Maximum in June 47 ° C
Precipitation	Max Annual 2060 mm/year in Uganda
Min Annual 0 mm/year in Egypt	
Mean Annual flow (Discharge) (km^3/year) at Aswan	84×109 m^3
Discharge/Unit area	28×10^3 m^3/km^2
Main Consumptive Water use	Agriculture

Source Hailu and Honours (2013a, b)

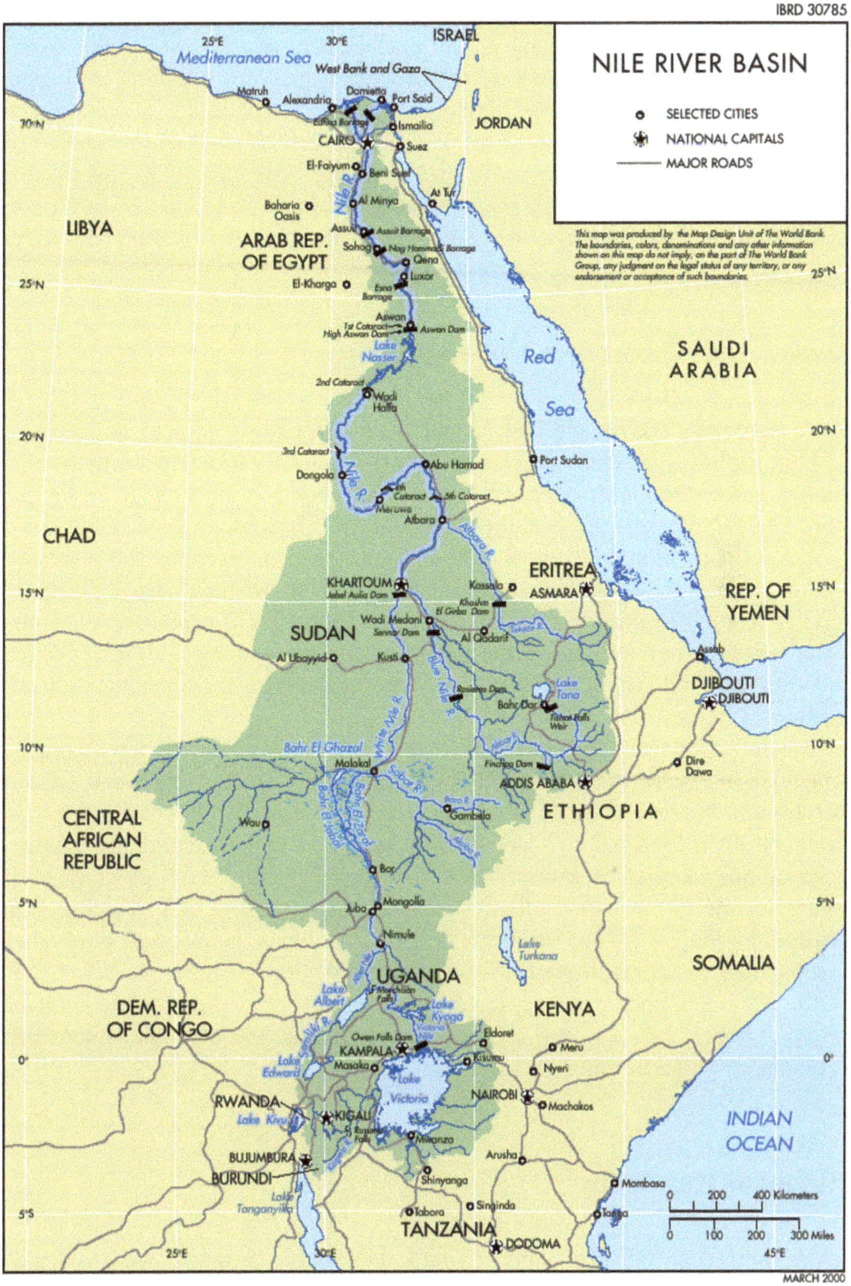

Fig. 3.4 The Nile Basin. *Source* www.nilebasin.org, 2009

Although it is true that all riparian states, including Egypt and Sudan, face critical water scarcity, development, and management issues (Tadesse 2007), this chapter, however, focuses primarily on Ethiopia's quest for economic and social development. The reason is that Ethiopia faces greater challenges of soil erosion, deforestation, and extreme famine, in spite of the contribution of more than 86% of the total of Nile Waters (Woldetsadik 2013). Another limitation is that due to space restrictions, it does not consider the environmental impact and human costs posed by hydro-dams, particularly on the Omo River Valley and on the Blue Nile River.

Puzzles and Theoretical Understandings

Hydro-hegemony (Egypt and Sudan) and Counter Hydro-Hegemony (Ethiopia).

What explains Ethiopia's bold approach (since 1991) to unilateral hydro-dam building and to the highly cooperative dialogues employed in challenging Egypt's hydro-hegemonic position over the Nile Waters? The change is puzzling. Most theories in International Relations (**IR**), from neo-realism (whereby states are seen as homogenous actors) to neo-liberal institutionalism (whereby states are understood in relation to institutions) (Brunnee and Toope 2013) explain conflicts and power relations from a single perspective of weak versus strong. Yet, such a casual theoretical analysis cannot sufficiently account for the counter-hegemonic influence of Ethiopia in the Nile River Basin.

Therefore, following the existing debates, the paper considers the theoretical framework of "hydro-hegemony" (Zeitoun and Warner 2012a, b) by **Mark Zeitoun and Jeroen Warner** and "counter-Hegemony" (Cascao 2011) conceptually and systematically developed by **Ana Elisa Cascao**. Both have strikingly captured **Gramsci-an** theory of "hegemony" and "counter-hegemony" both of which entail having legitimacy, consent, ideas, and knowledge (NBI 2010). Coming from **Gramsci-an** logic, **Cascao** argues that holding hydro-hegemonic power over others is about demonstrating leadership, but not employing coercive measures (NBI 2010). The importance of an approach to leadership focused on finding equitable solutions rather than the one focused on containing the problem is the main argument proposed in this book.

Ethiopian Hegemony

Before proceeded in explaining Egypt's hegemonic position and Ethiopia's counter-hegemonic position, it is important to briefly explain the concept of "hegemony". The concept of "hegemony" was first developed by **Antonio Gramsci** in a prison cell in fascist Italy, although he was unable to systematically elaborate the concept further due to his failing health (Bates 1975). Both hegemony and counter- hegemony theoretical concepts, thus, are influenced by **Gramsci-an** theory, and as well as by neo-**Gramsci-an** views (which are related to critical IR studies) (Cascao 2011). But what is "hegemony"? While there is a clear difference between "egemonia" (hegemony) and "dominante" (dominant), far too often the two are used interchangeably (Warner 2008a, b).

According to **Gramsci**, hegemony is a political power that flows from intellectual and moral leadership, authority or consensus as distinguished from armed force (Cascao 2011). Gramsci added to the definition of hegemony by stating that the political power obtained subsequently become the dominant hegemony by authority rather than by coercion (Gramsci 1971). Hegemony involves legitimacy and some form of understanding what would result from consent (Cascao 2011). However, whether Egypt will consent to new forms of legal and political legitimacy is yet to be seen.

Following **Gramsci-an** logic, hegemony is a process of achieving consent through newly proposed ideas that is understood as a sign of leadership rather than dominance (NBI 2010). Ideas and knowledge are powerful instruments that impact power relations in highly contested international basins (NBI 2010). As argued by **Cascao**, the production of knowledge is directly related to successful pursuance of material interest (Cascao 2011). For instance, the Grand Ethiopian Renaissance Dam is a material interest that was envisioned by the late **Prime Minster Meles Zenawi** and was also highly accepted by the multitudes. Regarding the latter development, **Gramsci** could say that the "Renaissance" is not only a political invocation but is also a necessary step forward for national progress to bring the peoples together (Gramsci 1971) domestically and regionally. Consequently, **Gramsci** argued that "…one should not count only on the material force which hegemony is understood gives in order to exercise an effective leadership (NBI 2010)." Instead, the cultural and social norms, as well as the ideas brought about by intellectuals are forms of hegemony (Abrahamsen 2013). Hegemony is understood here as the demonstration of effective leadership on the issue. This is also true for "organic intellectuals" (the working class peoples) where they can develop their own cultural intelligentsia to play a political role that is focused on the Party (Gramsci 1971). When a new common cultural bloc is formed between the traditional intellectuals and organic intellectuals the process of hegemony can be managed and maintained by the intellectuals (the powerful) as a whole (Cox 2013). Robert W. **Cox** argues that the intellectuals perform the function of developing and sustaining the mental images, technologies, and organizations which bind them together into a common identity or ideas (Cox 2013).

We can understand the above analysis in the following way that Meles Zenawi can be seen as the intellectual/political figure who managed to engage his fellow citizens (the masses) and the Ethiopian People Revolutionary Democratic Front (**EPRDF**) Party of the domestic and regional economic interests of the **GERD**. Zenawi's function as a visionary is still widely praised as hegemonic (dominant, but not coercive) by many Ethiopians and as well as by members of other riparian states. Thus, as **Gramsci** argued, the idea of having a common purpose behind a common issue is hegemonic, but the implementation of the idea by force is not (Gramsci 1971). **Gramsci** may also consider the emerging of new hydro-political relations in the Nile Basin under Meles Zenawi and its revolutionary party, the **EPRDF**, to be "the Modern Prince". The Modern Prince is, according to **Gramsci**, "the myth prince cannot be a real person, a concrete individual. It can only be an organism, a complex element of society in which a collective will, which has already been recognized and has to some extent asserted itself in action, begins to take concrete form (NBI 2010)". For **Gramsci**, the Modern Prince must have the collective or the political will to operate and be effective. Zenawi's revolutionary party have managed to maintain the political will as a collective political entity that has been able to govern for more than two decades by exerting national priorities, such as the question of the Nile waters. However, critics of Zenawi are well aware of how this single political party has managed to govern this long with an iron fist (Aalena and Tronvoll 2013). It may have been done through his leadership or coercive measures, or both. Despite skepticism, the legal, political, economic, and environmental questions and vision left behind by Zenawi over the Nile waters continue as stepping stones for further hydro-political relations in the Nile Basin.

Hydro-Hegemony (Egypt and Sudan)

The growing water conflict between upstream (Ethiopia) and downstream (Egypt and Sudan) riparian states calls for a deeper understanding of the asymmetric power relations in the Nile River Basin (Zeitoun and Warner 2012a, b). The theoretical framework of hydro-hegemony provides the analysis for power exertion in international watercourses by highlighting the role of the state. Even though the term is still loosely used, hydro-hegemony still captures the growing hydro-political tensions at most international river basins (Warner 2008a, b), especially in the Nile Basin. Egypt and Sudan are the hegemons (by authority or by dominance). Particularly, Egypt has been referred to as the most hegemonic downstream riparian state in the Nile Basin. The question then is: Why and how?

Historically, Egypt has exercised various hegemonic strategies to control the water resources of the Nile. One of them is through "resource capture (Cascao 2013b)". As defined by **Thomas F. Homer-Dixon**, resource capture is similar to rent seeking behavior, whereby some individuals or groups maximize their economic wellbeing by manipulating the rules of production (Homer-Dixon 1999). The motivation behind resource capture can be greed (monopolistic profits) or fear of rising scarcity that may be produced in the near future (NBI 2010). One of Egypt's contemporary (and historic) fears (as throughout history) is that the emerging capabilities of upstream states like Ethiopia to develop the source of the Nile River (Blue Nile) will subsequently reduce the regular amount of water flowing into Egypt. As outlined by **Cascao**, to date, there have been three historical hegemonic strategies used by Egypt to control the allocation of the Nile waters: the first phase was initiated by Muhammad Ali in the nineteenth century through an extensive expansion of irrigation infrastructure. The second phase was during the British colonial period, when several technological and political measures were pursued in order to launch a new hydraulic project intensifying Egyptian agricultural production. The third phase was the period of Arab nationalism, when Gamal Abdel-Naser initiated projects that culminated in the building of the Aswan High dam (Cascao 2013b).

The historical ideas and the knowledge that Egypt was able to construct around the Nile waters and then use for absolute unilateral Nile water resources development have led to contemporary complex and contested asymmetric power relations. Asymmetric power relations are a core component of the hydro-political relations in the Nile River Basin (Cascao 2013b) due to the unequal (highly contested) development and distribution of water resources between upstream and downstream riparian states. Following Gramsci's logic, Egypt's historical moves can be considered as a sign of leadership (within Egypt) rather than dominance by coercion, which subsequently led to unequal political power relations over the access to the Nile waters (outside Egypt). Moreover, Egypt's historical hegemony on the Nile water usage includes that an understanding by Egypt of its own contemporary and future interests and objectives (Gill and Law 1993).

In turn, the technical, legal, and political knowledge used to mobilize Egypt's historical interests over the Nile waters turned into, as **Zeitoun** and **Warner** argued, dominance defined as leadership buttressed by coercion (Zeitoun and Warner 2012a, b), which will be discussed further in the historical significance section of the research. Such unilateral domination was highly supported by the colonial master, mainly Britain, and then further reinforced during the proxy war by opposing superpowers Russia and the US. For instance, the British favored Egypt in the development of hydropower infrastructures while ignoring the needs of the upstream riparian states—most of which had been under British rule, except Ethiopia (Cascao 2013b). From the legal point of view, **Cascao** posits that the 1959 Agreement (which guarantees the allocation of 55.5 bcm per year to Egypt and 18.5 bcm per year to Sudan) reinforces the "natural and historical rights" to the Nile waters, which is the baseline for any possible negotiations with upstream states (Cascao 2013a). This means that the colonial agreement is neither up nor can be further negotiated, because for Egypt the Nile waters have always been classified as a matter of national security (Buzan et al. 1998). The invocation of potentially "national security" mobilizes and legitimizes all possible actions, even the use of force, to stop any claims that may jeopardize the status quo (Buzan et al. 1998). For instance, Egypt used "covert action" tactics in the past in support of various rebels within Ethiopia to weaken the political and hydraulic power of Ethiopia (Zeitoun and Warner 2012a, b).

The support of colonial Britain for Egypt was motivated by economic and geopolitical interest gains rather than by the security and stability of the Nile River Basin (Collins 1994). At the end, the colonial-legal pact made between the two has only benefited both countries. This was also true during the Cold War, when the Western superpowers strengthened Egyptian hydro-hegemony by supporting a unilateral construction of the Aswan dam (Cascao 2013b). Currently, Egyptian hydro-hegemonic order continues to benefiting through new but twisted kind of support vis-à-vis the growing geopolitical conflicts in the Middle East, namely via military aid from the US. As of June 2013, while tensions over water between Egypt and Ethiopia are growing by the day, the US has secretly approved its annual military aid supply worth USD 1.3 billion to Egypt (Chibarirwe 2013). The reasons for such a move could be multilayered. However, the continuous exertion of military, economic, and political power through external financial aid could be particularly detrimental to the overall human and economic security in one of the most fragile regions of Africa. The latter analysis will be statistically supported further that military aid influence does not sustain cooperation (Hailu and Honours 2013a, b).

Consequently, Egypt is a hydro-hegemon in terms of the capabilities it holds in the technical, political, financial, economic, and military spheres of influence compared to Ethiopia. Such capabilities are constructed around the historical and colonial-legal rights discourses regarding usage of the Nile waters (Hailu and Honours 2013a, b). Stefan Deconinck argues that both Egypt and Sudan appeal to their historical and legal rights (which were obtained under the British colonial rule) to expand their consumption of the Nile waters, while refusing to accept the needs of upstream riparian states (Deconinck 2013). However, this does not necessarily mean that upstream riparian states, such as Ethiopia, are powerless in the face of these hegemonic (by force) instruments (Cascao 2013b).

References

Aalena, L., & Tronvoll, K. (2013). The end of democracy? Curtailing political and civil rights in Ethiopia. *Review of African Political Economy, 36*(120), 193–207 (2009). Accessed May 21, 2013

Abrahamsen, R. (2005). Defines securitization following the Copenhagen School as a "political choice, a decision to conceptualize an issue in a particular ways…by an "us against them" distinction and a logic of existential-threat". In R. Abrahamsen (Ed.), *Blair's Africa: The politics of securitization and fear* (p. 60). Alternatives: Global, Local, and Political.

Abrahamsen, R. (2013). The theoretical aspects of hegemony between the material term used by Marx and the ideas and cultural norms and values described by Gramsci as seen as "hegemony". This is her interpretation following Gramsci's hegemony, June 24, 2013.

Arsano, Y. (2010). Institutional development and water management in the Ethiopian Nile Basin. In T. Tvedt (Ed.), The River Nile in the post-colonial age (p. 161). I. B. Tauris & Co. Ltd.

Arsano, Y. (2013). *Ethiopia and the Nile: Dilemmas of national and regional hydro politics* (pp. 82). Zurich: Swiss Federal Institute of Technology, 2007. http://archiv.css.ethz.ch/publications/Ethiopia_and_the_Nile.pdf. Accessed May 20, 2013

Bates, T. R. (1975). Gramsci and the theory of hegemony. *Journal of the History of Ideas, 36*(2), 351

Brunnee, J., & Toope, S. J. (2013). The Nile Basin regime: A role for law? *Water Resources Perspectives: Evaluation, Management and Policy*, 117 (2003): 95. Accessed May 5, 2013.

Buzan, B., et al. (1998). Define 'security' "is about survival…the invocation of security has been the key to legitimizing the use of force, but more generally it has opened the way for the state to mobilize…". In B. Buzan, O. Waever, & J. de Wilde (Ed.), *Security: A new framework for Analysis* (p. 21). London: Lynne Rienner Publishers, 1998.

Cascao, A. E. (2011). Power relations, conflict, and cooperation in the Eastern Nile River Basin. In S. S. Elmusa (Ed.), *The burden of resources—Oil and water in the Gulf and the Nile Basin* (p. 28). American University in Cairo Press, 2011.

Cascao, A. E. (2013a). Changing power Relations in the Nile River Basin: Unilateralism vs. cooperation, Water Alternatives, (2009): 245. Accessed on May 28, 2013, http://agris.fao.org/agris-search/search/display.do?f=2013/AV/AV201201457001457.xml;AV20120145788

Cascao, A. E. (2013b). Ethiopia—Challenges to Egyptian hegemony in the Nile Basin. *Water Policy, 10*(Supplement 2), 13–28 (2008). Accessed May 5, 2013. doi:10.2166/wp.2008.206

Chibarirwe, A. (2013). America Secretly Approves $1.3 Billion Worth of Military Aid to Egypt. The Trumpet.com, June 10, 2013. Accessed June 14, 2013. http://www.thetrumpet.com/article/10703.19.0.0/world/military/america-secretly-approves-13-billion-worth-of-military-aid-to-egypt

Collins, R. O. (1994). History, hydro politics, and the Nile: Nile control: myth or reality? In P. P. Howell & J. A. Allan (Ed.), *The Nile: sharing a scarce resource, A historical and technical review of water management and of economic and legal issues* (p. 111). Cambridge: Cambridge University Press, 1994.

Conway, D. (2009). A water balance model of the Upper Blue Nile in Ethiopia. *Hydrological Sciences Journal*, 267. Accessed June 18, 2013. http://dx.doi.org/10.1080/02626669709492024

Cox, R. W. (2013). Gramsci, hegemony and international relations: An essay in method. *Journal of International Studies*, 168. SAGE (1983). Accessed June 14, 2013. doi:10.1177/03058298830120020701

Deconinck, S. (2013). *Security as a threat to development: the geopolitics of water scarcity in the Nile River basin*. Focus Paper, 5 (2009). Accessed May 28, 2013. http://www.waternet.be/middle_east/security_development_nile.pdf

Egypt President Politicians Plotting Against Ethiopia's Dam English Subtitle. (2013). You Tube, Published on Jun 5, 2013. Accessed on June 8, 2013. http://www.youtube.com/watch?v=b_mpjU2qIdc

Gill, S., & Law, D. (1993). Global hegemony and the Structural power of capital. In S. Gill (Ed.), *Gramsci historical materialism and international relations* (p. 131). Cambridge: Cambridge University press, 1993.

Gramsci, A. (1971). *Selection from the prison notebooks* (p. 212) (Quentin Hoare and Geoffrey Nowell Smith, Trans.). London: Lawrence & Wishart

Hailu, D., & Honours B. A. (2013a). *Ethiopia's Emerging Counter-Hydro Hegemonic Influence: Changing the Tides of the Blue Nile Waters for an "equitable" Basin-Wide System (Cooperation and Integration)*. A Major Research Paper (MRP), Master of Arts in the Graduate School of Public and International Affairs (GSPIA), Ryerson University, 2011, Ottawa, Ontario, Canada, 2013.

Hailu, D., & Honours B. A. (2013b). *Ethiopia's emerging counter-hydro hegemonic influence: Changing the Tides of the Blue Nile Waters for an "equitable" Basin-Wide System (Cooperation and Integration)*. A Major Research Paper (MRP), Ryerson University, Ottawa, Ontario, Canada, 2013

Hassan, H. A., & Rasheedy, A. A. (2011). The Nile River and Egyptian foreign-policy interests. In K. G. Adar & N. A. Check (Ed.), Cooperative Diplomacy, Regional Stability and National Interests: The Nile River and Riparian States (p. 131). Africa Institute of South Africa.

Homer-Dixon, T. F. (1999). *Environment, scarcity, and violence* (p. 66). Princeton University Press.

ILA. (2013). The International Law Association as the Helsinki Rules, the UN Watercourses Convention and the Berlin Rules: Perspectives on International Water Law. *Water Resources Development,* 629. The World Bank, Washington DC, 2007. Accessed June 18, 2013. http://dx.doi.org/10.1080/07900620701488562

Kameri-Mbote, P. (2007). *Navigating peace: Water, conflict, and cooperation: Lessons from the Nile River Basin* (p. 1). Woodrow Wilson, International Center for Scholars.

Kandie, D. (1999). *Egypt and the hydro-politics of the Blue Nile River* (p. 142), Michigan State University Press. Accessed June 5, 2013.

Kandie, D. (2013). Egypt and the hydro-politics of the Blue Nile River. Michigan State University Press, 1999. Accessed June 5, 2013, http://dx.doi.org/10.1353/nas.2002.0002

Mason, S. A. (2004). *From conflict to cooperation in the Nile Basin* (p. 1) (Ph.D.). Swiss Federal Institute of Technology, eth Zurich.

Mumma Martinon, C. A. (2010). Nile Basin Initiative: A possibility of turning conflicts into opportunities. In B. Calas & C.A. Mumma Martinon (Eds.), *Shared waters, shared opportunities: Hydropolitics in East Africa* (p. 55) (Ph.D.). French Institute for Research in Africa (IFRA), 2010.

NBI. (2010). NBI Country Profile—Egypt's Policy. Accessed on June 10, 2013. http://www.nilebasin.org/newsite/index.php?option=com_content&view=category&layout=blog&id=35&Itemid=67&lang=en

NBI. (2012). Nile Basin Initiative, State of the River Nile Basin, Entebbe, Uganda.

Nile River. (2013). Nile Basin Initiative (NBI). Accessed June 10, 2013. http://www.nilebasin.org/newsite/index.php?option=com_content&view=category&layout=blog&id=36&Itemid=75&lang=en

Stratfor. (2013). The geopolitical impact of the Nile. Accessed July 8, 2013. http://www.stratfor.com/video/geopolitical-impact-nile

Tadesse, D. (2007). *The regional dimensions of Ethiopia's Economic and social development with special reference to the Nile River* (p. 32) (Ph.D.). Howard University.

Tafesse, T. (2011). Ethiopia's position on the Nile water agreements. In K. G. Adar & N. A. Check (Ed.), *Cooperative diplomacy, regional stability and national interests: The Nile River and Riparian States* (p. 78). Africa Institute of South Africa

UNDE. (2013). *Water Scarcity*, 2007. Accessed June 5, 2013. http://www.un.org/waterforlifedecade/scarcity.shtml

Warner, J. (2008a). Contested hydro hegemony: Hydraulic control and security in Turkey. *Water Alternatives*, 272

Warner, J. (2008b). Contested hydro hegemony: Hydraulic control and security in Turkey. *Water Alternatives*, 271–288.

Water Project. (2013). Poverty in Africa begins with A Lack of Clean water. The Water Project, 2013. Accessed June 8, 2013. http://thewaterproject.org/poverty.asp

Woldegebriel, E. G. (2013). Ethiopia hopes to reap benefits from eco-friendly rail projects. Thomson Reuters Foundation, Published 23 April 2013. Accessed June 10, 2013, http://www.trust.org/item/20130423142157-cofa6/

Woldetsadik, T. K. (2013). *International watercourses law in the Nile River Basin: Three states at a Crossroads* (p. xv). London: Routledge.

Yohannes, O. (2013). Hydro-politics in the Nile basin: in search of theory beyond realism and neo-liberalism. *Journal of Eastern African Studies*, 78 (2009). Accessed June 10, 2013. http://dx.doi.org/10.1080/17531050802682788

Zeitoun, M., & Warner, J. (2012a). Hydro-hegemony—A framework for analysis of trans-boundary water conflicts. *Water Policy, 8*, 443 (2006). Accessed December 12, 2012. http://dx.doi.org/10.2166/wp.2006.054

Zeitoun, M., & Warner, J. (2012b). Hydro-hegemony—A framework for analysis of trans-boundary water conflicts. *Water Policy, 8*, 453–460 (2006). Accessed December 12, 2012. doi:10.2166/wp.2006.054

Chapter 4
Dams and Barrages on River Channels

Abstract This chapter provides a brief overview background of dams and barrages on river channels, scope of the chapter, water and development, economic growth and development, competing uses of water, development and large dams, dams in the twentieth century, large dams as instruments of development, irrigation water supplied from large dams, problems associated with large dams, riverine ecosystems impacts, social consequences of large dams, understanding the large dams debate and social impacts and equity.

Keywords Scope of the chapter · Water and development · Economic growth and development · Competing uses of water · Development and large dams · Dams in the 20th century · Large dams as instruments of development · Irrigation water supplied from large dams · Problems associated with large dams · Riverine ecosystems impacts · Social consequences of large dams · Understanding the large dams debate · Social impacts and equity

Over 45,000 times in the last century, people took the decision to build a dam. Dams were built to provide water for irrigated agriculture, domestic or industrial use, to generate hydropower or help control floods. But dams also altered and diverted river flows, affecting existing rights and access to water, and resulting in significant impacts on livelihoods and the environment. Decisions to build dams are being contested increasingly as human knowledge and experience expand, as we develop new technologies, and as decision-making becomes more open, inclusive and transparent. The World Commission on Dams considers that the end of any dam project must be the sustainable improvement of human welfare. This means a significant advance of human development on a basis that is economically viable, socially equitable, and environmentally sustainable. If a large dam is the best way to achieve this goal it deserves our support. Where other options offer better solutions we should favor them over large dams. Thus the debate around dams challenges our view of how we develop and manage our water resources (Dams and development 2000).

© Springer International Publishing AG 2017

Y.M. Hamada, *The Grand Ethiopian Renaissance Dam, its Impact on Egyptian Agriculture and the Potential for Alleviating Water Scarcity*,
Environment & Policy 55, DOI 10.1007/978-3-319-54439-7_4

Background

Water has attracted the attention of political leaders at the highest level and has triggered a series of strategic global initiatives such as the World Commission on Water (World Commission on Water in the 21st Century 2000). The most powerful players in development have placed water at the top of their agendas. Why this sudden attention to a resource that was, after all, central to human needs long before the dawn of civilization? At one level, the water issue justifies the priority accorded it because demand for water is rapidly outstripping available supply in large parts of the world. As populations grow, and economic development leads to higher consumption, demand for water increases sharply putting intense pressure on available stocks. This can cause growing social tensions, or even lead to outright conflict. While commonly expressed predictions that future wars will largely be fought over water may be exaggerated, nobody doubts that access to both surface and ground water is an increasingly contentious issue.

Where co-operation gives way to unfair competition between different water uses or between communities and countries, a range of new issues come to the fore. These issues relate to the distribution of power and influence within societies and between countries. They concern the relative weight of the different factors that make up the decision-making mix. And they relate to how choices are made between available options.

Scope of the Chapter

The issues surrounding dams are the issues surrounding water, and how water-related decisions are made. There is little public controversy about the choice between an embankment dam or a gravity dam, or whether to use earth, concrete or rock-fill, possibly even over financing the development. The issues all relate to what the dam will do to river flow, to rights of access to water and river resources, to whether it will uproot existing human settlements, disrupt the culture and sources of livelihood of local communities and deplete and degrade environmental resources. Conflicts over dams are more than conflicts over water. They are conflicts over human development and life itself. If water is life, rivers are its arteries. Dams regulate or divert the flow through these arteries, affecting the lifeblood of humanity. The fact that they aim to do so in the interests of humankind only makes the decision to build a large dam more sensitive, one that will trigger a range of apprehensions, hopes and fears, both rational and irrational (World Commission on Water in the 21st Century 2000).

Water and Development

Today, around 3800 km^3 of freshwater is withdrawn annually from the world's lakes, rivers, and aquifers (Gleick 1998). This is twice the volume extracted 50 years ago. A growing population and a rising level of economic activity both increase human demand for water and water-related services. Development, technological change, income distribution, and lifestyles all affect the level of water demand.

World population has passed 6 billion. Although the annual increase probably peaked at about 87 million around 1990, the high proportion of young people in most developing countries means that global population will continue to increase significantly well into this century. Recent projections suggest a peak of between 7.3 billion and 10.7 billion around 2050 before total population begins to stabilize or fall (UN 1999; Raskin et al. 1998), (Medium projection in 2050). Predictions cannot be precise, because other dimensions of development such as access to health, education, income, birth control and other services influence the pace of population growth. Despite the massive investment in water resource management and particularly in dams, billions of children, women and men in rural areas lack access to the most basic water and sanitation services. Although problems of access are worst in rural areas, rapid urbanization is also increasing the demand for water-related services. In 1995, 46% of the world's population lived in urban areas. If current trends persist (and they may accelerate), that figure could reach 60% by the year 2030 and over 70% by 2050 (UN 1998), (In 30 years the largest population growth is projected to be in urban conglomerations; the 10 largest population centers in descending order of size would be Mumbai (Bombay), Shanghai, Chennai (Calcutta), Beijing, Delhi, Karachi, Tianjin, Metro Manila, Jakarta, and Dhaka). Most of this growth will take place in developing countries where an estimated 25–50% of urban inhabitants live in impoverished slums and squatter settlements. Lack of access to water in both rural and urban areas is not just a question of supply. It is partly due to inequitable access to existing supplies.

Urbanization implies an increasing concentration on water and energy demand in megacities, a switch to different lifestyles and consumption patterns, and a loss of productive agricultural land through urban expansion. It is a widely held view that lack of attention to development in rural areas is fuelling unsustainable forms of urban growth, shifting poverty from rural to urban areas, and contributing to rapidly growing demand for additional services. In heavily populated countries like China, India, and Indonesia many question the sustainability of the high rates of urbanization in megacities.

Economic Growth and Development

World economic activity has grown approximately fivefold since 1950 at a rate of about 4% per year (IPCC 1999), [Purchasing power parity (PPP) uses accounting exchange rates to compare or add country economic data and avoid distortions introduced by market exchange rates. In 1995 world Gross Product reached a figure of US$33.4 trillion at PPP]. The regional balance is changing, with significant growth in Asia over the past 25 years. At present, OECD countries continue to account for the largest share, amounting to 55% of world production at purchasing power parity, nearly 80% at market prices (Raskin et al. 1998), (OECD Data; 78% at market prices and 55% at purchasing power parity).

Economic growth has two implications for water demand. The first is that increased economic activity will increase the demand for water-related services regardless of whether the demand is satisfied by more efficient use of the existing supply, or by increasing the level of supply. The second is that both the development brought about by economic growth and the technological changes that accompany it will lead to structural changes in the pattern of goods and services that society produces, and in the way these services are provided. The water demand per unit of Gross Domestic Product (**GDP**) will depend on how these two components of economic growth are combined. Countries with the same product per capita but different production characteristics—for example, with large-scale irrigated agriculture or water intensive industries—may consume three or four times more water per dollar of GDP. This is evident when comparing the United States and Canada with Germany or France, or India with China.

Development and technological change can also save water per dollar of output. Between 1950 and 1990 the world economy grew by a factor of five while world water withdrawals only grew by a factor of 2 (Shiklomanov 1998). The last 50 years have seen a worldwide reduction in the amount of water per dollar of nonagricultural production as a result of improved technology, more recycling, enforcement of environmental standards, higher water prices, and industries moving away from natural resource intensive activities (Raskin et al. 1994). However, the water management practices and technologies that enable such advances are not widely available or promoted and are often absent where they are most needed. Despite the increasing number of options available, the total number of people without access to clean water is growing.

Economic activity and development affect income, income distribution, and lifestyles. These in turn affect the demand for water through changes in the level and composition of household consumption in areas such as diet, the use of household appliances and standards of sanitation. How much water is needed for one more person? Although climate and culture influence what constitutes an appropriate level of domestic water consumption, several international agencies and experts have proposed 50 L per person per day (or just over 18.25 m^3 a year) as an amount that covers basic human water requirements for drinking, sanitation, bathing, and food preparation. In 1990 over a billion people were below that level (Gleick 1998). On the other hand, households in developed countries and better-off households in developing cities use from 4 to 14 times the threshold of 50 L per person a day. Drastically lower average figures for domestic consumption in developing countries reflect not only different lifestyles and lower incomes, but also a huge backlog of unsatisfied demand. The lower average also masks extremely high consumption among better-off urban households and acute deprivation among rural and urban poor.

Competing Uses of Water

Water analysts foresee increased competition among water users in meeting the growing demand. They predict that competition will increase among the three largest water users in global terms. Agriculture accounts for about 67% of withdrawals, industry uses 19% and municipal and domestic uses account for 9% (Keller et al. 1998). Analysts foresee that these uses will continue to draw from the water needed to sustain natural systems. In dry climates, evaporation from large reservoirs, estimated at close to 5% of total water withdrawals, may also be a significant consumptive use of water (Shiklomanov 1998). Regional trends vary widely. Despite increasing urbanization in Africa, Asia and Latin America, agriculture is the dominant water user in these regions, accounting for approximately 85% of all water used. In all regions of the world except Oceania, domestic or household water consumption accounts for less than 20% of water use (Wright et al. 1998). In Africa, Central America and Asia, this is nearer to 5%. In the more developed regions of Europe and North America, industry is the major water consumer. The water use breakdown in a specific country influences where water demand management opportunities exist. For many water-stressed countries that are industrializing, and that have large urban populations, the present water crisis often revolves around industrial and domestic water supply and sanitation. This is evident in the significant trend over the past few decades toward diverting water from agricultural to municipal and industrial uses. As this is happening in the context of an overall increase in withdrawals, it will lead to increased competition for water, unless more efficient water use in both sectors accompanies the transition from agricultural- to industrial-based economies (Shiklomanov 1998).

There are additional challenges. To meet food requirements, water used in agriculture may have to increase 15–20% by 2025 even with improvements in irrigation efficiency and agronomic potential (Van Hofwegen and Svendsen 2000). In addition to increasing food production in the face of water stress, distribution, equitable access, purchasing power, and poverty are central issues in meeting food demand. Beyond competing human demands, water for nature is an essential consideration. The fresh water ecosystems that provide the livelihoods of the world's riverine communities and many other goods and services to our societies depend on water. Arresting, and where possible reversing, the accelerating trend to increasing degradation of many of the world's watersheds caused by human activity have emerged as an urgent priority.

Rainfall and other sources of freshwater (rivers, lakes, groundwater) are unevenly distributed around the world and are not always located where human water demand arises. Pollution threatens surface and ground water sources and may make them unfit for many uses, or require expensive treatment. Pollution is especially serious where ground water resources are overexploited and suffer from reduced natural rates of recharge due to deforestation, land use changes and urbanization. Fully one-third of the countries in water-stressed regions of the world are expected to face severe water shortages this century, and within these regions there are great disparities in access to fresh water. Not surprisingly, a significant number of less developed countries, including regions of India and China, are facing severe shortages, Countries are considered to be water stressed when the annual renewable water supply drops below 1700 m^3 per capita, and water scarce when it drops below 1000 m^3 (Raskin et al. 1994). With population growth, the number of countries in this category is increasing, and by 2025 there will be approximately 6.5 times as many people—a total of 3.5 billion—living in water-stressed countries (Raskin et al. 1994). The uneven distribution of water supply means that countries may have water surplus and water deficit regions. Not only surface water is under pressure. The growing rate of extraction of fresh water from rivers and lakes is matched by increasing extraction of ground water, with many aquifers now seriously depleted. The volume of ground water withdrawal, primarily for irrigation but also for municipal and industrial use, exceeds long-term recharge rates. In many parts of India, Pakistan and China, the water table is sinking at the rate of one to two meters a year (Brown et al. 1999). Climate change can also affect the seasonal distribution of rains and water availability. Studies and modeling exercises reviewed by the World Commission on Dams, including those by the Intergovernmental Panel on Climate Change (**IPCC**), strongly suggest that global warming trends could significantly increase the variability of weather patterns (WCD 2014). Results of World Commission on Dams could include a decline in rainfall in arid and semiarid regions, and an increase in seasonal variations in rainfall around the globe over the next 50–100 years.

Development and Large Dams

River basins are renowned as the cradles of civilization and cultural heritage. Ancient and modern communities alike have depended on rivers for livelihood, commerce, habitat and the sustaining ecological functions they provide. Throughout history alterations to rivers—natural or human generated—have affected riverine communities in one way or another. The earliest evidence of river engineering is the ruins of irrigation canals over eight thousand years old in Mesopotamia. Remains of water storage dams found in Jordan, Egypt and other parts of the Middle East date back to at least 3000 BC. Historical records suggest that the use of dams for irrigation and water supply became more widespread about a thousand years later. At that time, dams were built in the Mediterranean region, China and Miso America. Remains of earth embankment dams built for diverting water to large community reservoirs can still be found in Sri Lanka and Israel (Schnitter 1994; McCully 1996). The Dujiang irrigation project, which supplied 800,000 ha in China, is 2200 years old (Zhang 2000). Dams and aqueducts built by the Romans to supply drinking water and sewer systems for towns still exist today. The first use of dams for hydropower generation was around 1890. By 1900, several hundred large dams had been built in different parts of the world, mostly for water supply and irrigation.

Dams in the Twentieth Century

The last century saw a rapid increase in large dam building. By 1949 about 5000 large dams had been constructed worldwide, three-quarters of them in industrialized countries. By the end of the twentieth century, there were over 45,000 large dams in over 140 countries, compilation of estimates of the number of dams in the main dam building countries suggest there may be as many as 48,000 large dams, (ICOLD 1998). The period of economic growth following the Second World War saw a phenomenal rise in the global dam construction rate, lasting well into the 1970s and 1980s. At its peak, nearly 5000 large dams were built worldwide in the period from 1970 to 1975. The decline in the pace of dam building over the past two decades has been equally dramatic, especially in North America and Europe where most technically attractive sites are already developed. The average large dam today is about 35 years old. The top five dam building countries account for nearly 80% of all large dams worldwide. China alone has built around 22,000 large dams, or close to half the world's total number. Before 1949 it had only 22 large dams. Other countries among the top five dam building nations include the United States with over 6390 large dams; India with over 4000; and Spain and Japan with between 1000 and 1200 large dams each. Approximately two thirds of the world's existing large dams are in developing countries. Construction of large dams peaked in the 1970s in Europe and North America. Today most activity in these regions is focused on the management of existing dams, including rehabilitation, renovation and optimizing the operation of dams for multiple functions. An estimated 1700 large dams have been under construction in other parts of the world in the last few years. Of this total, 40% are reportedly being built in India. Since average construction periods generally range from 5 to 10 years, this indicates a worldwide annual average of some 160–320 new large dams per year (World Commission on Water in the 21st Century 2000).

The end of the twentieth century saw the emergence of another trend relating to large dams—decommissioning dams that no longer serve a useful purpose, are too expensive to maintain safely, or have unacceptable levels of impacts in today's view. Momentum for river restoration is accelerating in many countries, especially in the United States where nearly 500 dams, mainly relatively old, small dams have been decommissioned. Since 1998, the decommissioning rate for large dams has overtaken the rate of construction in the United States.

Experience in North America and in Europe shows that decommissioning dams has enabled the restoration of fisheries and riverine ecological processes. However, dam removals without proper studies and mitigation actions cause public concerns and environmental problems. These include negative impacts on downstream aquatic life due to a sudden flush of the sediments accumulated in the reservoir. Where there has been industrial or mining activity upstream, these sediments may be contaminated with toxic substances. Another option is to open the floodgates of dams where this is possible, either as a form of decommissioning, or to allow natural river flows and unrestricted fish passage at critical times of the year. While decommissioning efforts in the United States and France have received public support thus far, there may be local opposition where changes in the flow and water levels affect services previously provided by the dam, or where development has taken place around the reservoir and downstream. There is comparatively little experience with the removal of larger dams. The bigger dam, the more problems decommissioning or removal are likely to face, and the more expensive they are likely to be. More studies are needed to address the costs, benefits, and impacts of decommissioning as the stock of dams ages and choices must be made between refurbishing and decommissioning (World Commission on Water in the 21st Century 2000).

Large Dams as Instruments of Development

Dams have been promoted as an important means of meeting perceived needs for water and energy services and as long-term, strategic investments with the ability to deliver multiple benefits. Some of these additional benefits are typical of all large public infrastructure projects, while others are unique to dams and specific to particular projects. Regional development, job creation, and fostering an industry base with export capability are most often cited as additional considerations for building large dams. Other goals include creating income from export earnings, either through direct sales of electricity, or by selling cash crops or processed products from electricity intensive industry such as aluminum refining. Water-rich countries such as Canada, Norway, Brazil, and parts of Russia have developed large dams for hydropower generation where suitable sites were available. Governments in semiarid countries such as South Africa, Australia and Spain have tended to build dams with large storage capacity to match water demand with stored supply, and for security against the risk of drought. For example, in Spain—one of the top five dam building countries—rainfall is highly variable between seasons and from year to year (WCD 2014).

In East and Southeast Asia, during the monsoon season, rivers swell to over 10 times the dry season flow. In these settings dams have been constructed to capture and store water during wet seasons for release during dry seasons. Large dams require significant financial investments. Estimates suggest a worldwide investment of at least two trillion US dollars in the construction of large dams over the last century. During the 1990s, an estimated $32–46 billion was spent annually on large dams, four-fifths of it in developing countries. Of the $22–31 billion invested in dams each year in developing countries, about four-fifths was financed directly by the public sector (WCD 2014).

Today the world's large dams regulate, store and divert water from rivers for agricultural production, human and industrial use in towns and cities, electricity generation, and flood control. Dams have been constructed to a lesser extent to improve river transportation and, once created for other purposes, the reservoirs of many large dams have been used for recreation, tourism, and aquaculture. The majority of large dams in Africa and Asia are for irrigation, though large dams are more often than not multipurpose. There is growing interest in dams for flood protection and in pumped storage dams for power generation to meet peak demand in Asia. Single-purpose hydropower dams are most common in Europe and South America, whereas single-purpose water supply projects dominate in Austral-Asia. North America has a relatively even spread of large dam functions. All other potential purposes, including recreation and navigation, were found in less than 5% of projects. Overall, the proportional share of irrigation dams and multipurpose dams has been increasing over the last 20 years, while the share of hydropower dams has been decreasing (WCD 2014).

Irrigation Water Supplied from Large Dams

Irrigation is the single largest consumptive use of fresh water in the world today. It is linked to food production and food security. About one-fifth of the world's agricultural land is irrigated, and irrigated agriculture accounts for about 40% of the world's agricultural production (WCD 2014). The total area irrigated expanded dramatically during the first years of the green revolution in the 1960s, increasing yields and bringing down food prices. From 1970 to 1982, global growth in the irrigated area slowed to 2% a year. In the post green revolution period between 1982 and 1994 it declined to an annual average of 1.3% (Cosgrove and Rijsberman 1999).

Half of the world's large dams were built exclusively or primarily for irrigation, and an estimated 30–40% of the 268 million hectares of irrigated lands worldwide rely on dams. Discounting conjunctive use of ground water and surface water, dams are estimated to contribute to at most 12–16% of world food production. Ground water irrigation typically has higher yields, for a given amount of water, than surface water irrigation because of better control of the resource at farm level. Four countries—China, India, United States, and Pakistan account for more than 50% of the world's total irrigated area.

Dams supply the water for almost 100% of irrigated production in Egypt—most coming from the Aswan High Dam—while in Nepal and Bangladesh dams provide only 1% of irrigation water. In the two countries with the largest irrigated areas—India and China—official statistics suggest large dams supply approximately 30–35% of irrigation water, with the balance coming mainly from ground water sources. There is some controversy in calculating the percentage of food production attributable to dams, and particularly on the methods used to account for conjunctive use of surface and ground water, World Commission on Dams (2000) India dams supply the water for almost 30% of irrigated land with suggested an alternative figure of 10%. Unsustainable irrigation practices have affected more than a fifth of the world's irrigated area in arid and semiarid regions. As a result of World Commission on Dams, soil salinity and waterlogging either make agriculture impossible, or limit yields and the types of crop that can be grown. In other regions, overuse of tube wells has depleted ground water aquifers, lowering water tables and making extraction increasingly expensive and especially difficult for smallholders. The absence of effective policies on conjunctive use of ground water and surface water resources is one of the most important concerns.

Problems Associated with Large Dams

While dams have contributed to economic growth in the twentieth century, the services they provide have come at a cost. This section gives an overview of the problems associated with large dams.

Large dams have fragmented and transformed the world's rivers. The World Resources Institute (**WRI**) found that at least one large dam modifies 46% of the world's 106 primary watersheds (Revenga et al. 1998). The extent to which river flows have been changed varies around the world. The United States and the European Union regulate the flow of 60–65% of the rivers in their territories, though the amount varies from basin to basin. Spain's 53 km^3 of storage behind large dams regulates 40% of its river flow, varying from 71% in the Ebro river basin, to 11% in the basins on the Galicia coast (Berger et al. 2000). In Asia, just fewer than half the rivers that are regulated have more than one large dam (ICOLD 1998). The modification of river flows on Trans Boundary Rivers has particular implications. There are 261 watersheds that cross the political boundaries of two or more countries (Wolf 2000). These basins cover about 45% of the earth's land surface, account for about 80% of global river flow and affect about 40% of the world's population (Yoffe et al. 2003). Issues and concerns facing riparian countries in these basins range from water quality to flow volumes. The aggregate storage capacity of large dams, based on dam design, is about 6000 km^3 (Lecornu 1998). Assuming that half the design storage is achieved in reality, the aggregate real storage capacity of large dams globally is similar to total freshwater withdrawals estimated at around 3800 km^3, a considerable portion of the aggregate storage of large dams may be for nonconsumptive hydropower releases. Analysis of all large dams registered in the ICOLD (1998), but excluding single-purpose hydropower dams, shows 4373 km^3 of designed storage capacity. This data set excludes many of the large dams in China, estimated to provide 451 km^3 of storage (IJHD 1999). An estimated 0.5–1% of the total fresh water storage capacity of existing dams is lost each year to sedimentation in both large and small reservoirs worldwide (Keller et al. 2000). This means that 25% of the world's existing fresh water storage capacity may be lost in the next 25–50 years in the absence of measures to control sedimentation. This loss would mostly be in developing countries and regions, which have higher sedimentation rates.

Riverine Ecosystems Impacts

Growing threats to the ecological integrity of the world's watersheds come from rising populations, water pollution, deforestation, withdrawals of water for irrigation and municipal water supply and the regulation of water flows resulting from the construction of large dams. Among the many factors leading to the degradation of watershed ecosystems, dams are the main physical threat, fragmenting and transforming aquatic and terrestrial ecosystems with a range of effects that vary in duration, scale and degree of reversibility. The watersheds of the world are the habitat of 40% of the world's fish species, and provide many ecosystem functions ranging from nutrient recycling and water purification to soil replenishment and flood control. At least 20% of the world's more than 9000 fresh water fish species have become extinct, threatened or endangered in recent years (Revenga et al. 1998).

Fish are a critical source of animal protein for more than 1 billion people. In Africa, fish protein is 21% of the total animal protein in the diet, and in Asia it is 28%. While rivers supply about 6% of fish protein consumed by humans globally, it is often 100% of the supply for many inland riverine communities. Ecosystem transformations do not only occur in the upper, lower and mid-reaches of watersheds, they also impact on river estuaries, which are frequently complex ecosystems. Closing the mouths of major rivers, salt intrusion, destruction of mangroves and loss of wetlands are among the many issues at stake (WCD 2014).

Social Consequences of Large Dams

While many have benefited from the services large dams provide, their construction and operation have led to many significant, negative social and human impacts. The adversely affected populations include directly displaced families, host communities where families are resettled, and riverine communities, especially those downstream of dams, whose livelihood and access to resources are affected in varying degrees by altered river flows and ecosystem fragmentation. More broadly, whole societies have lost access to natural resources and cultural heritage that were submerged by reservoirs or rivers transformed by dams. The construction of large dams has led to the displacement of some 40–80 million people worldwide (Revenga et al. 1998). Many of them have not been resettled or received adequate compensation, if any (Fox and Brown 1998). Between 1986 and 1993, an estimated 4 million people were displaced annually by an average of 300 large dams starting construction each year (World Bank 1996). The scale and range of social issues encountered in the river basins altered by large dam construction vary from region to region. The number of people deriving their livelihood directly from the river and its ecosystem, and the overall population density in the river basin, gives an indication of the potential impacts. The world's two most populous countries—China and India, have built around 57% of the world's large dams—and account for the largest number of people displaced. In the late 1980s China officially recognized some 10.2 million people as 'reservoir resettles', though other sources suggest the figure may be substantially higher (Jing 1999). Large dams and reservoirs already built in the Yangtze basin alone have led to the relocation of at least 10 million people. In India, estimates of the total number of people displaced due to large dams vary from 16 to 38 million people (WCD 2014). Resettlement caused by large dams has been a significant part of total resettlement for all public infrastructure development. In China, large dams are estimated to have displaced an estimated 27% of all people displaced by development projects (the total includes people displaced by urban expansion, roads and bridges) (ADB 2000). In India the figure is 77% (this total excludes people displaced by urban development) (Fernandes and Paranjpye 1997). Among World Bank funded projects involving displacement, dams and reservoirs accounted for 63% of people displaced (World Bank 1996). These figures are only estimates and do not include people displaced by other aspects of the projects such as canals, powerhouses, project infrastructure, and associated compensatory measures such as bioreserves.

Understanding the Large Dams Debate

The huge growth of dam building in the twentieth century took place against a backdrop of tremendous political, economic and technical transformations—while the world's population grew from 1.65 billion in 1900, to 6 billion by the end of the century. The last three decades especially have seen dramatic and wide-reaching changes in perceptions of development and concepts of interdependence with other people and with nature. This dynamic of change is also redefining the roles that governments, civil society and the private sector play in water and energy resource development planning. And while the wider transformations have spurred the debate on large dams that debate it has become a major catalyst of change. By looking at the genesis of opposition to large dams from the perspective of social and environmental movements, before considering the main issues in the debate that the meeting in Gland Switzerland handed down to the World Commission on Dams.

The huge growth of dam building in the River Nile took place. Figure 4.1 shows placement of early dams on the Nile. Table 4.1 shows most Nile countries will build more dams on the Nile.

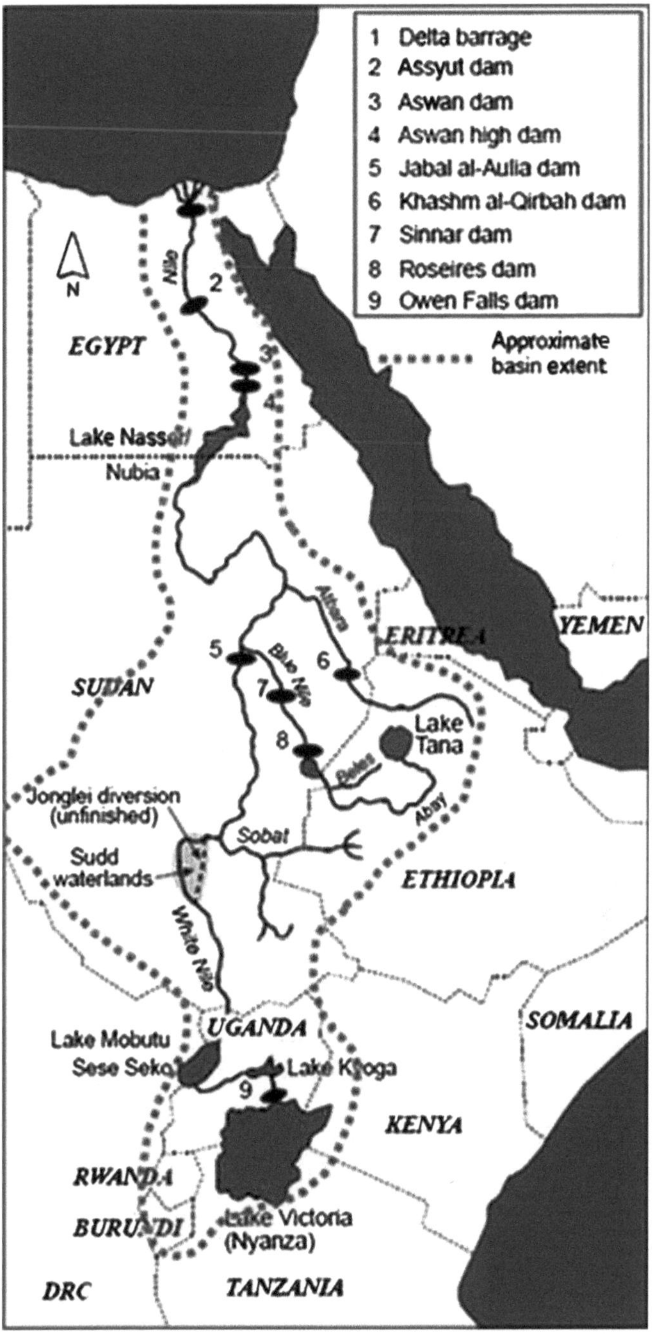

Fig. 4.1 Placement of early dams on the Nile. *Sources* Nicol (2003)

Table 4.1 Major dams and barrages finished, unfinished and planned in the Nile Basin

Country	Name of dam	River	Year completed (or to be started)	Power (MW)	Storage (m^3)	Contractor
1900–1970s postindependence						
Egypt	Assiut barrage	Main Nile	1902	Irrigation		
Egypt	Esna barrage	Main Nile	1908	Irrigation		
Egypt	Nag-Hamady barrage	Main Nile	1930	Irrigation		
Egypt	Old Aswan Dam	Main Nile	1933		4,50,000	
Egypt	High Aswan Dam	Main Nile	1970	2100	11,10,000	
Ethiopia	Tis-Abay	Lake Tana	1953	12		
Sudan	Sennar	Blue Nile	1925	48	0.93	
Sudan	Jebel Aulia	White Nile	1937	18		
Sudan	Khashm El Gibra	Atbara	1964	35	1.3	
Sudan	Roseires	Blue Nile	1966	60	2.386	
Uganda	Owen/Nalubaale	White Nile	1954	180	0.23	
1970 to present						
Ethiopia	Tekeze 5	Tekeze	2009–2010	300	9.2	
Sudan	Merowe	Main Nile	2009–2010	350	12	
Ethiopia	Finchaa	Finchaa	1971–2013	134	1050	
Ethiopia	Chara Chara	Blue Nile	2000	84	9126	
Ethiopia	Koga	Blue Nile	2008	Irrigation	80	
Ethiopia	Tana Beles	Blue Nile	2011		460	
Kenya	Sondu	Miriu Victoria	2007	60	1.1	Japan
Uganda	Kiira/extension	White Nile	1993–2000		200	

(continued)

Table 4.1 (continued)

Country	Name of dam	River	Year completed (or to be started)	Power (MW)	Storage (m^3)	Contractor
Under construction (date gives completion date)						
Sudan	Roseires	Blue Nile	2013	Multiheightening		National
Sudan	Burdana	Setit/Atbara		135		China/Kuwait
Sudan	Rumela	Atbara		135		China/Kuwait
Sudan	Shiraik	Main Nile		300		
Ethiopia	FAN Finchaa		2011			China/Italy
Ethiopia	Tekeze II	Tekeze	2020			
Ethiopia	Megech	Abay		Irrigation		Multinational
Ethiopia	Ribb	Abay	2011			
Ethiopia	Grand	Blue Nile	2017		5250	China/Millennium Italy
Rwanda	Nyabarongo	Nyabarongo	2011		27.5	Australia/India
Uganda	Bujagali	White Nile	2011		250	Italy
Dams proposed (date gives potential start date)						
DRC	Semliki	Semliki				
Ethiopia	Jema	Jema				
Ethiopia	Karadobi	Blue Nile	2023	1600		ENSAP
Ethiopia	Border	Blue Nile	2026	1400		ENSAP
Ethiopia	Mabil	Blue Nile	2021			
Ethiopia	Beko Abo	Blue Nile	2000			ENSAP
Ethiopia	Mendaya	Blue Nile	2030	1700		ENSAP
Ethiopia	Chemoda/Yeda	Chemoga	2015	278		China

(continued)

Table 4.1 (continued)

Country	Name of dam	River	Year completed (or to be started)	Power (MW)	Storage (m^3)	Contractor
Five dams proposed Yeda rivers						
Ethiopia	Baro I	Sobat				
Ethiopia	Baro II	Sobat				
Sudan	Nimule	Nile				
Sudan	Dal-1	Nile		400		
Sudan	Kajbar	Nile		300		
Sudan	South Bedden	Bahr el Jebel				Italy/NBI
Sudan	South Shukoli	Bahr el Jebel				Italy/NBI
Sudan	South Lakki	Bahr el Jebel				Italy/NBI
Sudan	South Fula	Bahr el Jebel				Italy/NBI
Uganda	Isimba	White Nile	2015	87		
Uganda	Kalagala	White Nile	2011	300		India
Uganda	Karuma	White Nile	2017	200		
Uganda	Murchison	White Nile		600		
Uganda	Ayago North	White Nile	2018	304		
Uganda	Ayago South	White Nile		234		
Uganda	15	small Kagera				
Run-of-the river						
Rwanda	Kikagate	Kagera	2016	10		
Rwanda	Nyabarongo	Kagera	2012	27		
Rwanda	Rusumo I & II	Kagera	2012	60		NELSAP
Tanzania						

(continued)

Table 4.1 (continued)

Country	Name of dam	River	Year completed (or to be started)	Power (MW)	Storage (m^3)	Contractor
Burundi						
Kenya	Goronga	Mara				
Kenya	Machove	Mara				
Kenya	Kilgoris	Mara				
Kenya	EwasoNgiro	Mara	2012	180		UK

Sources Ofcansky and Berry (1991), Nicol (2003), Scudder (2005), Dams and Agriculture in Africa (2007), McCartney (2007), World Bank (2007), UNEP (2008), African Dams Briefing (2010), Dams and Hydropower (2010), Verhoeven et al. (2011)

Note ENSAP = Eastern Nile Subsidiary Action Program

Social Impacts and Equity

This includes both the scale of the impacts and the distribution of costs, benefits and impacts, including those borne by relocated families, host communities where families are resettled and riverine communities affected by the change in river flows and access to resources. Social issues go beyond equity in the distribution of benefits and impacts and relate to fundamental rights. They include: the burden placed on indigenous peoples and ethnic minorities and the degree of recognition of their distinctive status; the impact on gender and basic human rights and the loss of livelihood and health impacts in rural areas. Dealing with the legacy left by forced relocation under both authoritarian and democratic regimes and the need to allocate responsibility for redress are other issues in the debate. An issue that has caused a great deal of tension in the past concerns the basis on which trade-offs, such as the potential benefits to many at the cost of hardship for a few, are invoked and decided (WCD 2014).

Dam critics point to:

- the need for more sustainable and appropriate alternatives to dams.
- the imperative for improved transparency, accountability and public participation in the planning of water and energy projects.
- the importance of prior project approval by potentially affected groups.
- the need for protecting and promoting the rights of potentially affected peoples, and for setting in place measures to reduce inequities.
- the necessity of reparation measures to address the legacy of unfulfilled commitments and unresolved problems.

Dam proponents underline:

- the evolution and change in practices over time
- the recognized need for social and environmental concerns to be elevated to the same level as safety concerns. (They believe these environmental and social costs must be avoided, or minimized and properly mitigated when they occur)
- the importance of ensuring that affected people are better-off as a result of dam development, and of considering them as Water, Development and Large Dams The Report of the World Commission on Dams shareholders, partners, and therefore project beneficiaries
- the imperative of participatory decision-making
- the need to promote the principles of equity, efficiency and economic viability
- the importance of balancing the need for development with the requirement of ensuring environmental sustainability.

References

ADB. (2000). Resources and financial management, special theme: Water in the 21st century, institutional document. 306-8370 (print), April, 2000.
African Dams Briefing & Map. (2010). *International Rivers' Africa Program.* Berkeley: Caledonia.
Berger, T., Brigl, M., Herrmann, J. M., Vielhauer, V., Luckow, B., Schlondorff, D., et al. (2000). The apoptosis mediator mDAP-3 is a novel member of a conserved family of mitochondrial proteins. *Journal of Cell Science, 113*(Pt 20), 3603–3612.
Brown, L. R., Gardner, G. T., & Brian, H. (1999). *Beyond Malthus, nineteen dimensions of the population challenge.* The World Watch Environmental Alert Series, Linda Starke, Series Editor, www.worldwatch.org. © Copyright 1999 World Watch Institute.
Brown, P., Beekes. W, & Verhoeven, P. (2010). *Corporate governance, accounting and finance: A review* (pp. 96–172), Version of Record online: 11 NOV 2010| DOI:10.1111/j.1467-629X.2010.00385.x
Cosgrove, W. J., & Rijsberman, F. R. (1999). *World water vision: Work plan 1999–2000.* Paris: World Water Vision Unit.
Dams and Agriculture in Africa. (2007). McCartney, World Bank.
Dams and Development. (2000). *A new framework for decision-making, The report of the world commission on dams.* London and Sterling, VA: Earth scans Publications Ltd, November 2000. http://www.earthscan.co.uk
Dams and Hydropower. (2010). The International Journal on Hydropower and Dams is a bi-monthly publication.
Fox, J., & David, B. L. (1998). *The struggle for accountability: The World Bank.* NGOs and Grassroots Movements, Cambridge: MIT Press.
Gleick, P. H. (1998). The world's water 1998–1999: The biennial report on freshwater resources.
ICOLD. (1998). The World Register of Dams (WRD) has been created by ICOLD in 1958.
IPCC. (1999) The Intergovernmental Panel on Climate Change, Geneva, Friday, 4 June 1999.
Jing, J. (1999). Contributing paper for WCD thematic review 1.3 Displacement, p. 2.
Keller, A. A., Sakthivadivel, R., & Seckler, D. (1998). *Water scarcity and the role of storage in development.* International Water Management Institute: Colombo, Sri Lanka.
Keller, M. B., McCullough, J. P., Klein, D. N., Arnow, B., Dunner, D. L., Gelenberg, A. J. (2000). A comparison of nefazodone, the cognitive behavioral-analysis system of psychotherapy, and their combination for the treatment of chronic depression. *The New England Journal of Medicine 342*(20), 1462–1470.
Lecornu. (1998). Flood control dams, irrigation dams, domestic and industrial water supply dams, recreation dams, fish breeding dams, and navigation dams.
McCully, P. (1996). Environmental effects of hydrological alterations, used the ratio of large: Small dams in the United States.
Nicol, A. (2003). The Nile: moving beyond cooperation. Unesco, www.unesco.org/"aterlwwap/pccp!casc_stlldies.shtml, accessed March 2011.
Ofcansky, T. P., & Berry, L. V. B. (1991). Ethiopia: A country study—American Memory—Library of Congress.
Raskin, I., Gleba, D., & Smith, R. (1998). *Dams and development: A new framework for decision.* United Nations Environment Programmed, Dams and Development Project.
Raskin, I., Nanda-Kumar, P. B. A., Dushenkov, S., & Salt, D. E. (1994). Bio concentration of heavy metals by plants. *Current Opinion Biotechnology, 5,* 285–290.
Revenga, C., et al. (1998). The WRI's 1998 study of the conditions of the major watersheds in the world focused on 150 watersheds, representing 55% of the world's land area.
Schnitter, N. J. (1994). A history of dams, how dams vary and why it matters for the emerging science of dam removal. *Oxford Journals Science & Mathematics Bioscience, 52*(8), 659–668.
Schnitter, N. J. (2003). *A history of dams: the useful pyramids.* Washington, D.C.: Georgetown University Press.

Scudder, T. (2005).*The future of large dams: Dealing with social, environmental, institutional and political costs*, London.
Shiklomanov, I. A. (1998). *World water resources and their use a joint*. International Hydrological Programmed (IHP) of UNESCO.
UN. (1998). Secretary-general's reports submitted to the security council in 1998, Security Council Resolutions, Resolutions adopted by the Security Council in 1998.
UN. (1999). Secretary-general's reports submitted to the security council in 1999, Security Council Resolutions, Resolutions adopted by the Security Council in 1999.
UNEP. (2008). An overview of the State of the world's fresh and marine waters—2nd Edition.
Van Hofwegen, P., & Svendsen, M. (2000). *Water for agriculture: Irrigation economics in international. Case study: Irrigation management transfer on the Alto Río Lerma*. Mexico.
Verhoeven, H. (2011). Black Gold for Blue Gold?: Sudan's Oil, Ethiopia's Water and Regional Integration, Africa Programme Briefing Paper. Chatham House, London. www.chathamhouse.org.uk/research/africa/papers, accessed March 2011.
Walter, F., & Vijay P. (1997). Rehabilitation policy and the law in India. Indian Social Institute, New Delhi. Ministry of Agriculture, 15. 4.
WCD. (2014). Worst-case distance, How to analyze high yield and debug, improve the design in Virtuoso ADE, Virtuoso Analog Design Environment.
Wolf, A. T. (2000). Contributing paper for WCD thematic review V.3 River Basins.
World Bank. (1996). World Development Report 1996: From plan to market, World Development Report examines the transition of countries with alternative systems of centrally planned economies back to a market orientation.
World Bank. (2007). The International Bank for Reconstruction and Development. The World Bank.
World Commission on Water in the 21st Century. (2000). Other examples of strategic initiatives in water include the United Nations comprehensive assessment of freshwater resources of the world, the United Nations Food and Agriculture Organization Aqua stat programed, and the world water vision conferences.
Wright, E. F., Coates, k. D., Can ham, C. D., & Bartemucci, P. (1998). Species variability in growth response to light across climatic regions in northwestern British Columbia. *Canadian Journal of Forest Research*, *28*, 871–886.
Yoffe, S., Wolf, A. T., Giordano, M. (2003). Conflict and cooperation over international freshwater resources: Indicators of basins at risk. *Journal of the American Water Resources Association*.
Zhang. (2000). World commission on dams regional consultation paper.

Chapter 5
Effects of Dams and Barrages on River Channels

Abstract This chapter provides a brief overview of effects of dams and barrages as: effects of dams on river channels, effects of gravel mining on river channels, effects of denudation on river channels, channel incision, bed coarsening and loss of spawning gravels, gravel replenishment below dams, sediment sluicing and pass-through from reservoirs, channel narrowing and fine sediment accumulation below dams and coastal erosion.

Keywords Scope of the chapter · Effects of dams on river channels · Effects of gravel mining on river channels · Effects of denudation on river channels · Channel incision · Bed coarsening and loss of spawning gravels · Gravel replenishment below dams · Sediment sluicing and pass-through from reservoirs · Channel narrowing and fine sediment accumulation below dams and coastal erosion

Dams and diversions are constructed and operated for a wide variety of purposes including residential, commercial, and agricultural water supply; flood and/or debris control; and hydropower production. Regardless of their purpose, all dams trap sediment to some degree and most alter the flood peaks and seasonal distribution of flows, thereby profoundly changing the character and functioning of rivers. By changing flow regime and sediment load, dams can produce adjustments in alluvial channels, the nature of which depends upon the characteristics of the original and altered flow regimes and sediment loads. Dams disrupt the longitudinal continuity of the river system and interrupt the action of the conveyor belt of sediment transport. Upstream of the dam, all bed load sediment and all or part of the suspended load (depending upon the reservoir capacity relative to inflow) (Brune 1953) is deposited in the quiet water of the reservoir (reducing reservoir capacity) and upstream of the reservoir in reaches influenced by backwater. Downstream, water released from the dam possesses the energy to move sediment, but has little or no sediment load. This clear water released from the dam is often referred to as hungry water, because the excess energy is typically expended on erosion of the channel bed and banks for some years following dam construction, resulting in incision (down cutting of the bed) and coarsening of the bed material until equilibrium is reached and the material

© Springer International Publishing AG 2017

Y.M. Hamada, *The Grand Ethiopian Renaissance Dam, its Impact on Egyptian Agriculture and the Potential for Alleviating Water Scarcity*,
Environment & Policy 55, DOI 10.1007/978-3-319-54439-7_5

cannot be moved by the flows. Reservoirs also may reduce flood peaks downstream, potentially reducing the effects of hungry water, inducing channel shrinking, or allowing fine sediments to accumulate in the bed.

Background

The World Commission on Dams considers that the end of any dam project must be the sustainable improvement of human welfare. This means a significant advance of human development on a basis that is economically viable, socially equitable, and environmentally sustainable. If a large dam is the best way to achieve this goal, it deserves our support. Where other options offer better solutions we should favor them over large dams. Thus the debate around dams challenges our view of how we develop and manage our water resources (DAD 2000).

Water has attracted the attention of political leaders at the highest level and has triggered a series of strategic global initiatives such as the World Commission on Water (DAD 2000). The most powerful players in development have placed water at the top of their agendas. Why this sudden attention to a resource that was, after all, central to human needs long before the dawn of civilization?

At one level, the water issue justifies the priority accorded it because demand for water is rapidly outstripping available supply in large parts of the world. As populations grow, and economic development leads to higher consumption, demand for water increases sharply putting intense pressure on available stocks. This can cause growing social tensions, or even lead to outright conflict. While commonly expressed predictions that future wars will largely be fought over water may be exaggerated, nobody doubts that access to both surface and ground water is an increasingly contentious issue (DAD 2000).

Where cooperation gives way to unfair competition between different water uses or between communities and countries, a range of new issues come to the fore. These issues relate to the distribution of power and influence within societies and between countries. They concern the relative weight of the different factors that make up the decision-making mix. And they relate to how choices are made between available options (DAD 2000).

Scope of the Chapter

This chapter focuses on the rivers transport sediment from eroding uplands to depositional areas near sea level. If the continuity of sediment transport is interrupted by dams or removal of sediment from the channel by gravel mining, the flow may become sediment-starved (hungry water) and prone to erode the channel bed and banks, producing channel incision (down cutting), coarsening of bed material, and loss of spawning gravels for salmon and trout (as smaller gravels are transported without replacement from upstream). Damming and mining have reduced sediment delivery from rivers to many coastal areas, leading to accelerated beach erosion. Sand and gravel are mined for construction aggregate from river channel and floodplains. In-channel mining commonly causes incision, which may propagate up- and downstream of the mine, undermining bridges, inducing channel instability, and lowering alluvial water tables. Floodplain gravel pits have the potential to become wildlife habitat upon reclamation, but may be captured by the active channel and thereby become in stream pits. Management of sand and gravel in rivers must be done on a regional basis, restoring the continuity of sediment transport where possible and encouraging alternatives to river-derived aggregate sources (Kondolf 1997a, b).

Effects of Dams

Comprehensive management of gravel and sand in river systems should be based on a recognition of the natural flow of sediment through the drainage network and the nature of impacts (to ecological resources and to infrastructure) likely to occur when the continuity of sediment is disrupted. A sediment budget should be developed for present and historical conditions as a fundamental basis for evaluation of these impacts, many of which are cumulative in nature (Kondolf 1997a, b). Lot of effects happened on rivers and its coasts as.

Effects of Dams on River Channels

Dams and diversions are constructed and operated for a wide variety of purposes including residential, commercial, and agricultural water supply; flood and/or debris control; and hydropower production. Regardless of their purpose, all dams trap sediment to some degree and most alter the flood peaks and seasonal distribution of flows, thereby profoundly changing the character and functioning of rivers. By changing flow regime and sediment load, dams can produce adjustments in alluvial channels, the nature of which depends upon the characteristics of the original and altered flow regimes and sediment loads. Dams disrupt the longitudinal continuity of the river system and interrupt the action of the conveyor belt of sediment transport. Upstream of the dam, all bed load sediment and all or part of the suspended load (depending upon the reservoir capacity relative to inflow) (Brune 1953) is deposited in the quiet water of the reservoir (reducing reservoir capacity) and upstream of the reservoir in reaches influenced by backwater. Downstream, water released from the dam possesses the energy to move sediment, but has little or no sediment load. This clear water released from the dam is often referred to as hungry water, because the excess energy is typically expended on erosion of the channel bed and banks for some years following dam construction, resulting in incision (down cutting of the bed) and coarsening of the bed material until equilibrium is reached and the material cannot be moved by the flows. Reservoirs also may reduce flood peaks downstream, potentially reducing the effects of hungry water, inducing channel shrinking, or allowing fine sediments to accumulate in the bed.

Effects of Gravel Mining on River Channels

As waters flow from high elevation to sea level, their potential energy is converted to other forms as they sculpt the landscape, developing complex channel networks and a variety of associated habitats. Rivers accomplish their geomorphic work using excess energy above that required to simply move water from one point on the landscape to another. In natural channels, the excess energy of rivers is dissipated in many ways: in turbulence at steps in the river profile, in the frictional resistance of cobbles and boulders, vegetation along the bank, in bends, in irregularities of the channel bed and banks, and in sediment transport. The transport of sand- and gravel-sized sediment is particularly important in determining channel form, and a reduction in the supply of these sediments may induce channel changes. The supply of sand and gravel may be the result of many factors, including changes in land use, vegetation, climate, and tectonic activity. This book is concerned specifically with the response of river channels to a reduction in the supply of these sediments by dams and gravel mining.

Sediment is transported mostly as suspended load: clay, silt, and sand held aloft in the water column by turbulence, in contrast to bed load: sand, gravel, cobbles, and boulders transported by rolling, sliding, and bouncing along the bed (Leopold et al. 1964). Bed load ranges from a few percent of total load in lowland rivers to perhaps 15% in mountain rivers (Collins and Dunne 1990) to over 60% in some arid catchments (Schick and Lekach 1993). Although a relatively small part of the total sediment load, the arrangement of bed load sediments constitutes the architecture of sand- and gravel-bed channels. Moreover, gravel and cobbles have tremendous ecological importance, as habitat for benthic macro-invertebrates and as spawning habitat for salmon and trout (Kondolf and Wolman 1993).

The rate of sediment transport typically increases as a power function of flow; that is, a doubling of flow typically produces more than a doubling in sediment transport (Richards 1982), and most sediment transport occurs during floods.

Effects of Denudation on River Channels

Viewed over a long term, run-off erodes the land surface, and the river network carries the erosional products from each basin. The rates of denudation, or lowering of the land by erosion, range widely. The Appalachian Mountains of North America are being denuded about 0.01 mm/year (Leopold et al. 1964), the central Sierra Nevada of California about 0.1 mm/year (Kondolf and Matthews 1993), the Southern Alps of New Zealand about 11 mm/year (Griffiths and McSaveney 1983), and the southern Central Range of Taiwan over 20 mm/year (Hwang 1994). The idealized watershed can be divided into three zones: that of erosion or sediment production (steep, rapidly eroding headwaters), transport (through which sediment is moved more or less without net gain or loss), and deposition (Schumm 1977). The river channel in the transport reach can be viewed as a conveyor belt, which transports the erosional products downstream to the ultimate depositional sites below sea level. The size of sediment typically changes along the length of the river system from gravel, cobbles, and boulders in steep upper reaches to sands and silts in low-gradient downstream reaches, reflecting diminution in size by water has always been a highly contentious issue in regional and international affairs, because of its importance for political and socioeconomic stability of nations. Without access to water, the maximization and protection of human and economic development is difficult, if not impossible. The management of this scarce resource in equitable and sustainable terms is currently an important debate within the Nile Basin-wide region, addressing the issues of poverty, population growth, soil erosion, environmental degradation and climate change. The Nile River waters are highly important to the evolving conflict between eleven riparian states: Kenya, Sudan, South Sudan, Burundi, Eritrea (observer status), Republic of Congo, Rwanda, Uganda, Tanzania, Ethiopia and Egypt, weathering and abrasion, as well as sorting of sizes by flowing water.

Transport of sediment through the catchment and along the length of the river system is continuous. Increased erosion in the upper reaches of the catchment can affect the river environment many miles downstream (and for years or decades) as the increased sediment loads propagate downstream through the river network. On Redwood Creek in Redwood National Park, California, the world's tallest trees are threatened with bank erosion caused by channel aggradation (building up of sediment in the channel), which in turn was caused by clear-cutting of timber on steep slopes in the upper part of the catchment (Madej and Ozaki 1996).

Along the river channel conveyor belt, channel forms (such as gravel bars) may appear stable, but the grains of which they are composed may be replaced annually or biannually by new sediment from upstream. Similarly, the sediments that make up the river floodplain (the valley flat adjacent to the channel) are typically mobile on a time scale of decades or centuries. The floodplain acts as a storage reservoir for sediments transported in the channel, alternately storing sediments by deposition and releasing sediment to the channel by bank erosion. For example, the Carmel River, California, is flanked by flat surfaces (terraces) that step up from the river. The lowest terrace is the channel of sand and gravel deposited by the 1911 flood, but the surface now stands about 4 m above the present, incised channel (Kondolf and Curry 1986). By 1960, the terrace had been subdivided for low-density housing, despite the recent origin of the land and the potential for future shifts in channel position.

A river channel and floodplain are dynamic features that constitute a single hydrologic and geomorphic unit characterized by frequent transfers of water and sediment between the two components. The failure to appreciate the integral connection between floodplain and channel underlies many environmental problems in river management today.

Channel Incision

Incision below dams is most pronounced in rivers with fine-grained bed materials and where impacts on flood peaks are relatively minor (Williams and Wolman 1984). The magnitude of incision depends upon the reservoir operation, channel characteristics, bed material size, and the sequence of flood events following dam closure. For example, the easily eroded sand bed channel of the Colorado River below Davis Dam, Arizona, has incised up to 6 m, despite substantial reductions in peak flows (Williams and Wolman 1984). In contrast, the Mokelumne River below Comanche Dam in California has experienced such a dramatic reduction in flood regime (and consequent reduction in sediment transport capacity) that no incision has been documented and gravels are reported to have become compacted and immobile (FERC 1993). Reduction in bed load sediment supply can induce a change in channel pattern, as occurred on Stony Creek, a tributary to the Sacramento River 200 km north of San Francisco. Since the closure of Black Butte Dam in 1963, the formerly braided channel has adopted a single-thread meandering pattern, incised, and migrated laterally, eroding enough bed load sediment to compensate for about 20% of the bed load now trapped by Black Butte Dam on an annual average basis (Kondolf and Swanson 1993).

Bed Coarsening and Loss of Spawning Gravels

Channel erosion below dams is frequently accompanied by a change in particle size on the bed, as gravels and finer materials are winnowed from the bed and transported downstream, leaving an armor layer, a coarse lag deposit of large gravel, cobbles, or boulders. Development of an armor layer is an adjustment by the river to changed conditions because the larger particles are less easily mobilized by the hungry water flows below the dam. The armor layer may continue to coarsen until the material is no longer capable of being moved by the reservoir releases or spills, thereby limiting the ultimate depth of incision (Williams and Wolman 1984; Dietrich and others 1989).

The increase in particle size can threaten the success of spawning by salmonids (salmon and trout), which use freshwater gravels to incubate their eggs. The female uses abrupt upward jerks of her tail to excavate a small pit in the gravel bed, in which she deposits her eggs and the male releases his milt. The female then loosens gravels from the bed upstream to cover the eggs and fill the pit. The completed nests (redds) constitute incubation environments with intragravel flow of water past the eggs and relative protection from predation. The size of gravel that can be moved to create a red depends on the size of the fish, ranging in median diameter from about 15 mm for small trout to about 50 mm for large salmon (Kondolf and Wolman 1993).

Below dams, the bed may coarsen to such an extent that the fish can no longer move the gravel. The Upper Sacramento River, California, was once the site of extensive spawning by Chinook salmon (Oncorhynchus tshawytscha), but massive extraction of gravel from the riverbed, combined with trapping of bed load sediment behind Shasta Dam upstream and release of hungry water, has resulted in coarsening of the bed such that spawning habitat has been virtually eliminated in the reach (Parfitt and Buer 1980). Figure 5.1 shows Keswick Dam and the channel of the Sacramento River downstream. The availability of spawning gravels can also be reduced by incision below dams when formerly submerged gravel beds are isolated as terrace or floodplain deposits. Encroaching vegetation can also stabilize banks and further reduce gravel recruitment for redds (Hazel et al. 1976).

Gravel Replenishment Below Dams

Gravels were being artificially added to enhance available spawning gravel supply below dams on at least 13 rivers in California as of 1992 (Kondolf and Matthews 1993). The largest of these efforts is on the Upper Sacramento River, where from 1979 to 2000 over US$22 million will have been spent importing gravel (derived mostly from gravel mines on tributaries) into the river channel (Denton 1991). Figure 5.2 shows gravel replenishment to the Sacramento River below Keswick Dam. While these projects can provide short-term habitat, the amount of gravel added is but a small fraction of the bed load deficit below Shasta Dam, and gravels placed in the main river have washed out during high flows, requiring continued addition of more imported gravel (California Department of Water Resources 1995).

On the Merced, Tuolumne, and Stanislaus rivers in California, a total of ten sites were excavated and back-filled with smaller gravel to create spawning habitat for Chinook salmon from 1990 to 1994. However, the gravel sizes imported were mobile at high flows that could be expected to occur every 1.5–4.0 years, and subsequent channel surveys have demonstrated that imported gravels have washed out (Kondolf et al. 1996a, b). On the border between France and Germany, a series of hydroelectric dams was constructed on the River Rhine (progressing downstream) after 1950, the last of which (the Barrage Iffezheim) was completed in the 1970s. To address the sediment deficit problem downstream of Iffezheim, an annual average of 170,000 tons of gravel (the exact amount depending on the magnitude of the year's run-off) is added to the river. Figure 5.3 shows Barge artificially feeding gravel into the River Rhine downstream of the Barrage Iffezheim. This approach has proved successful in preventing further incision of the riverbed downstream (Kuhl 1992). It is worth noting that the quantity of gravel added each year is not equivalent to the unregulated sediment load of the Rhine; the river's capacity to transport sediment has also been reduced because the peak discharges have been reduced by reservoir regulation. The amount of sediment added satisfies the transport capacity of the existing channel, which has been highly altered for navigation and hydroelectric generation.

Fig. 5.1 Keswick Dam and the channel of the Sacramento River downstream. *Source* Kondolf (1997a, b)

Fig. 5.2 Gravel replenishment to the Sacramento River below Keswick Dam. *Source* Kondolf (1997a, b)

Sediment Sluicing and Pass-Through from Reservoirs

The downstream consequences of interrupting the flux of sand and gravel transport would argue for designing systems to pass sediment through reservoirs and thereby reestablish the continuity of sediment transport.

To date, most such efforts have been undertaken to solve problems with reservoir sedimentation, particularly deposits of sediment at tunnel intakes and outlet structures, rather than to solve bed load sediment supply problems downstream. These efforts have been most common in regions with high sediment yields such as Asia (Sen and Srivastava 1995; Chongshan et al. 1995; Hassanzadeh 1995). Small diversion dams(such as those used to divert water in run-of-the-river hydroelectric generating projects) in steep V-shaped canyons have the greatest potential to pass sediment. Because of their small size, these reservoirs (or fore bays) can easily be drawn down so that the river's gradient and velocity are maintained through the dam at high flow. Large-capacity, low-level outlets are required to pass the incoming flow and sediment load.

If low-level outlets are open at high flow and the reservoir is drawn down, a small reservoir behaves essentially as a reach of river, passing inflowing sediment through the dam outlets. In such a sediment pass-through approach, the sediment is delivered to downstream reaches in essentially the same concentration and seasonal flood flows as prevailed in the pre-dam regime. This approach was employed at the old Aswan Dam on the River Nile and on the Bhatgurk Reservoir on the Yeluard River in India (Stevens 1936). Similarly, on the River Inn in Austria and Germany,

Fig. 5.3 Barge artificially feeding gravel into the River Rhine downstream of the Barrage Iffezheim. *Source* Kondolf (1997a, b)

floodwaters with high suspended loads are passed through a series of hydropower reservoirs in a channel along the reservoir bottom confined by training walls (Hack 1986; Westrich et al. 1992). If topographic conditions are suitable, sediment-laden floodwater may be routed around a reservoir in a diversion tunnel or permitted to pass through the length of the reservoir as a density current vented through a bottom sluice on the dam (Morris 1993). The Nan-Hwa Reservoir in Taiwan was designed with a smaller upstream fore bay from which sediment is flushed into a diversion tunnel, allowing only relatively clear water to pass into the main reservoir downstream (Morris 1993).

If sediment is permitted to accumulate in the reservoir and subsequently discharged as a pulse (sediment sluicing), the abrupt increase in sediment load may alter substrate and aquatic habitat conditions downstream of the dam. The most severe effects are likely to occur when sediment accumulated over the flood season is discharged during base flow (by opening the outlet pipe or sluice gates and permitting the reservoir to draw down sufficiently to resuspend sediment and move bed load), when the river's transporting capacity is inadequate to move the increased load. On the Kern River, the Southern California Edison Company (an electric utility) obtained agency permission to sluice sand from Democrat Dam in 1986, anticipating that the sand would be washed from the channel the subsequent winter. Figure 5.4 shows sand deposited in the bed of the Kern River as a result of sluicing from Democrat Dam in 1986. However, several years of drought ensued, and the sand remained within the channel until high flows in 1992 (Dan Christenson, California Department of Fish and Game, Kernville, personal communication 1992).

Fig. 5.4 Sand deposited in the bed of the Kern River as a result of sluicing from Democrat Dam in 1986. *Source* Kondolf (1997a, b)

On those dams larger than small diversion structures, the sediment accumulated around the outlet is usually silt and clay, which can be deleterious to aquatic habitat and water quality (Bjornn and Reiser 1991). Opening of the low-level outlet on Los Padres Dam on the Carmel River, California, released silt and clay, which resulted in a large fish kill in 1980 (Buel 1980). The dam operator has since been required to use a suction dredge to maintain the outlet (D. Dettman, Monterey Peninsula Water Management District, personal communication 1990). On the Dan River in Danville, Virginia, toxicity testing is required during sluicing of fine sediments from School field Dam (FERC 1995). Accidental sluices have also occurred during maintenance or repair work, sometimes resulting in substantial cleanup operations for the dam operators (Ramey and Beck 1990; Kondolf 1995).

Less serious effects are likely when the sediment pulse is released during high flows, which will have elevated suspended loads, but which can typically disperse the sediment for some distance downstream. The Jansanpei Reservoir in Taiwan is operated to provide power for processing only from November to April. The reservoir is left empty with open low-level outlets for the first 2 months of the rainy season (May and June), so sediments accumulated over the months of July–April can be flushed by the first high flows of the season before storing water in the latter part of the rainy season (Hwang 1994).

At present, sediment pass-through is not commonly done in North America, probably because of the limited capacity of many low-level outlets and because of concern that debris may become stuck in the outlets, making them impossible to close later, and making diversions impossible during the rest of the wet season until flows drop sufficiently to fix the outlets. These concerns can probably be addressed with engineering solutions, such as trash racks upstream of the outlet and redundancies in gate structures on the low-level outlet. Large reservoirs cannot be drawn down sufficiently to transport sediment through their length to the outlet works, for such a drawdown would eliminate carryover storage from year to year, an important benefit from large reservoirs.

In most reservoirs in the United States, sediment is simply permitted to accumulate. Active management of sediment in reservoirs has been rare, largely because the long-term costs of reservoir storage lost to sedimentation have not been incorporated into decision-making and planning for reservoirs. Most good reservoir sites are already occupied by reservoirs, and where suitable replacement reservoir sites exist, the current cost of replacement storage (about US\$3/m^3 in California) is considerably higher than original storage costs. Mechanical removal is prohibitively expensive in all but small reservoirs, with costs of \$15–\$50/m^3 cited for the Feather River in California (Kondolf 1995).

Channel Narrowing and Fine Sediment Accumulation Below Dams

While many reservoirs reduce flood peaks, the degree of reduction varies considerably depending upon reservoir size and operation. The larger the reservoir capacity relative to river flow and the greater the flood pool available during a given flood, the greater the reduction in peak floods. Flood control reservoirs typically contain larger floods than reservoirs operated solely for water supply. Downstream of the reservoir, encroachment of riparian vegetation into parts of the active channel may occur in response to a reduction in annual flood scour and sediment deposition (Williams and Wolman 1984). Channel narrowing has been greatest below reservoirs that are large enough to contain the river's largest floods. In some cases, fine sediment delivered to the river channel by tributaries accumulates in spawning gravels because the reservoir-reduced floods are inadequate to flush the riverbed clean.

On the Trinity River, California, construction of Trinity Dam in 1960 reduced the 2-year flow from 450 to 9 m^3/s. As a result of this dramatic change in flood regime, encroachment of vegetation and deposition of sediment has narrowed the channel to 20–60% of its pre-dam width (Wilcock et al. 1996). Accumulation of tributary-derived decomposed granitic sand in the bed of the Trinity River has led to a decline of invertebrate and salmonid spawning habitat (Fredericksen and Associates 1980). Experimental, controlled releases were made in 1991, 1992, 1995, 1993, and 1996 to determine the flows required to flush the sand from the gravels (Wilcock et al. 1996).

Such flushing flows increasingly have been proposed for reaches downstream of reservoirs to remove fine sediments accumulated on the bed and to scour the bed frequently enough to prevent encroachment of riparian vegetation and narrowing of the active channel (Reiser et al. 1989). The objectives of flushing flows have not always been clearly specified, nor have potential conflicts always been recognized. For example, a discharge that mobilizes the channel bed to flush interstitial fine sediment will often produce comparable transport rates of sand and gravel, eliminating the selective transport of sand needed to reduce the fine sediment content in the bed, and resulting in a net loss of gravel from the reach given its lack of supply from upstream (Kondolf and Wilcock 1996).

Coastal Erosion

Beaches serve to dissipate wave action and protect coastal cliffs. Sand may be supplied to beaches from headland erosion, river transport, and offshore sources. If sand supply is reduced through a reduction in sediment delivery from rivers and streams, the beach may become undernourished, shrink, and cliff erosion may be accelerated. This process by which beaches are reduced or maintained can be thought of in terms of a sediment balance between sources of sediment (rivers and headland erosion), the rate of long shore transport along the coast, and sediment sinks (such as loss to deeper water offshore) (Inman 1976). Along the coast of southern California, discrete coastal cells can be identified; each with distinct sediment sources (sediment delivery from river mouths) and sinks (losses to submarine canyons). For example, for the Oceanside littoral cell, the contribution from sediment sources (Santa Margarita, San Luis Rey, and San Diegueno rivers and San Mateo and San Juan creeks) was estimated, under natural conditions, at 209,000 m^3/year, roughly balancing the long shore transport rate of 194,000 m^3/year and the loss into the La Jolla submarine canyon of 200,000 m^3/year (Inman 1985).

The supply of sediment to beaches from rivers can be reduced by dams because dams trap sediment and because large dams typically reduce the magnitude of floods, which transport the majority of sediment (Jenkins et al. 1988). In southern California rivers, most sediment transport occurs during infrequent floods (Brownlie and Taylor 1981), but it is these energetic events that flood control dams are constructed to prevent. On the San Luis Rey River, one of the principal sources of sediment for the Oceanside littoral cell, Henshaw Dam reduced suspended sediment yield by 6 million tons, total sand and gravel yield by 2 million tons (Brownlie and Taylor 1981).

Ironically, by trapping sediment and reducing peak flows, the flood control dams meant to reduce property damage along rivers contribute to property damage along the coast by eliminating sediment supply to the protective beaches. For the rivers contributing sediment to the Oceanside littoral cell as a whole, sediment from about 40% of the catchment area is now cut off by dams. Because the rate of long shore transport (a function of wave energy striking the coast) is unchanged, the result has been a sediment deficit, loss of beach sand, and accelerated coastal erosion (Inman 1985).

The effects of sediment trapping by dams have been exacerbated in combination with other effects such as channelization and in stream sand and gravel mining (discussed below). Although sluicing sediment from reservoirs has been considered in the Los Angeles Basin, passing sediment through urban flood control channels could cause a number of problems, including decreasing channel capacity (Potter 1985). "Beach nourishment" with imported sediment dredged from reservoirs and harbors has been implemented along many beaches in southern California (Inman 1976; Allayaud 1985; Everts 1985). In some cases, sand is transported to critical locations on the coast via truck or slurry pipelines. The high costs of transportation, sorting for the proper size fractions, and cleaning contaminated dredged material, as well as the difficulty in securing a stable supply of material make these options infeasible in some places (Inman 1976).

To integrate considerations of fluvial sediment supply in the maintenance of coastal beaches into the existing legal framework, a system of "sand rights", analogous to water rights, has been proposed (Stone and Kaufman 1985).

References

Allayaud, W. K. (1985). Innovations in non-structural solutions to preventing coastal damage. In J. McGrath (Ed.), *California's Battered Coast, Proceedings from a Conference on Coastal Erosion* (pp. 260–290). California Coastal Commission.

Bjornn, T. L., & Reiser, D. W. (1991). Habitat requirements of salmonids in streams. In *Influences of forest and rangeland management on salmonid fishes and their habitats* (pp. 83–138). American Fisheries Society Special Publication 19.

Brownlie, W. R., & Taylor, B. D. (1981). Sediment management for southern California mountains, coastal plain, and shoreline. Part C. Coastal sediment delivery by major rivers in southern California. Report. 17-C, Environmental Quality Lab, California Institute of Technology, Pasadena.

Brune, G. M. (1953). The trap efficiency of reservoirs. *Transactions of the American Geophysical Union, 34,* 407–418.

Buel, B. (1980). Effects of Los Padres Reservoir silt release. Unpublished memo. Monterey Peninsula Water Management District, Monterey, California.

California Department of Water Resources. (1995). Sacramento River gravel restoration phase II study: a plan for continued spawning gravel replenishment between Keswick Dam and clear Creek. Technical Information Record TIR ND 95 1. California Department of Water Resources, Northern District, Red Bluff, California.

Chongshan, Z., Jianguo, W., & Quigmei, L. (1995). Experiment study of approach for sediment removed from reservoirs. In *Proceedings of Sixth International Symposium on River Sedimentation* (pp. 149–154). New Delhi, India.

Collins, B., & Dunne, T. (1990). Fluvial geomorphology and river gravel mining: a guide for planners, case studies included. California Division of Mines and Geology Special Publication 98. Sacramento.

DAD. (2000). *Dams and development, a new framework for decision-making, the report of the world commission on dams.* London and Sterling, VA: Earth scan Publications Ltd.

Denton, D. N. (1991). Sacramento River gravel restoration progress report. Unpublished report, California Department of Water Resources, Red Bluff, California.

Dietrich, W. E., Kirchner, J. W., Ikeda, H., & Iseya, F. (1989). Sediment supply and development of coarse surface layer in gravel bedded rivers. *Nature, 340,* 215–217.

Everts, C. H. (1985). Effects of small protective devices on beaches. In J McGrath (Ed.), *California's Battered Coast, Proceedings from a Conference on Coastal Erosion* (pp. 127–138). California Coastal Commission.

FERC (Federal Energy Regulatory Commission). (1993). Final environmental impact statement, propose modifications to the Lower Mokelumne River Project, California, FERC Project No. 2916-004. Washington, DC.

FERC (Federal Energy Regulatory Commission). (1995). Order amending and approving sediment flushing plan, STS Hydropower Limited and Dan River Incorporated, FERC Project No. 2411-012. Washington, DC.

Fredericksen, K., & Associates. (1980). Proposed Trinity River Basin fish and wildlife management program. Unpublished report to US Water and Power Resources Service (now the US Bureau of Reclamation).

Griffiths, G. A., & McSaveney, M. J. (1983). Hydrology of a basin with extreme rainfalls—Cropp River, New Zealand. *New Zealand Journal of Science, 26,* 293–306.

Hack, H. P. (1986). Design and calculation of reservoirs of run of river stations incorporating sedimentation. In W. Bechteler (Ed.), *Transport of Suspended Solids in Open Channels, Proceedings of Euromech 192* (pp. 107–112). Munich, Germany, 11–15 June 1985.

Hassanzadeh, Y. (1995). The removal of reservoir sediment. *Water International, 20,* 151–154.

Hazel, C., Herrera, S., Rectenwald, H., & Ives, J. (1976). Assessment of effects of altered stream flow characteristics on fish and wildlife. Part B: California case studies. Report by Jones and Stokes, Inc. to US Department of Interior, Fish and Wildlife Service.

Hwang, J. S. (1994). *A study of the sustainable water resources system in Taiwan considering the problems of reservoir desilting*. Taichung City, Taiwan: Taiwan Provincial Water Conservancy Bureau.

Inman, D. L. (1976). Man's impact on the California coastal zone. Summary report to California Department of Navigation and Ocean Development, Sacramento.

Inman, D. L. (1985). Budget of sand in southern California; river discharge vs. cliff erosion. In McGrath J (Ed.), *California's Battered Coast, Proceedings from a Conference on Coastal Erosion* (pp. 10–15). California Coastal Commission.

Jenkins, S. A., Inman, D. L., & Skelly, D. W. (1988). Impact of dam building on the California coastal zone. *California Waterfront Age September*.

Kondolf, G. M. (1995). Managing bedload sediments in regulated rivers: Examples from California, USA. *Geophysical Monograph, 89,* 165–176.

Kondolf, G. M. (1997a). Hungry water: Effects of dams and gravel mining on river channels. *Environment Management, 21*(4), 533–551.

Kondolf, G. M. (1997b). Effects of dams and gravel mining on river channels. *Environmental Management, 21*(4), 533–551 (Springer, New York Inc.).

Kondolf, G. M., & Curry, R. R. (1986). Channel erosion along the Carmel River, Monterey County, California. *Earth Surface Processes and Landforms, 11,* 307–319.

Kondolf, G. M., & Matthews, W. V. G. (1993). Management of coarse sediment in regulated rivers of California. Report No. 80. University of California Water Resources Center, Davis, California.

Kondolf, G. M., & Swanson, M. L. (1993). Channel adjustments to reservoir construction and in stream gravel mining, Stony Creek, California. *Environmental Geology and Water Science, 21,* 256–269.

Kondolf, G. M., Vick, J. C., & Ramirez, T. M. (1996a). Salmon spawning habitat rehabilitation in the Merced, Tuolumne, and Stanislaus Rivers, California: An evaluation of project planning and performance. Report No. 90, University of California Water Resources Center, Davis, California.

Kondolf, G. M., Vick, J. C., & Ramirez, T. M. (1996b). Salmon spawning habitat rehabilitation on the Merced River, California: An evaluation of project planning and performance. *Transactions of the American Fisheries Society, 125,* 899–912.

Kondolf, G. M., & Wilcock, P. R. (1996). The flushing flow problem: Defining and evaluating objectives. *Water Resources Research, 32*(8), 2589–2599.

Kondolf, G. M., & Wolman, M. G. (1993). The sizes of salmonid spawning gravels. *Water Resources Research, 29,* 2275–2285.

Kuhl, D. (1992). 14 years of artificial grain feeding in the Rhine downstream the barrage Iffezheim. In *Proceedings 5th International Symposium on River Sedimentation* (pp. 1121–1129). Karlsruhe, Germany.

Leopold, L. B., Wolman, M. G., & Miller, J. P. (1964). *Fluvial processes in geomorphology* (p. 522). San Francisco: W.H. Freeman.

Madej, M. A., & Ozaki, V. (1996). Channel response to sediment wave propagation and movement, Redwood Creek, California, USA. *Earth Surface Processes and Landforms, 21,* 911–927.

Morris, G. L. (1993). A global perspective of sediment control measures in reservoirs. In S. Fan & G. L. Morris (Eds.), *Notes on sediment management in reservoirs: National and international perspectives*. Washington, DC: US Federal Energy Regulatory Commission.

Parfitt, D., & Buer, K. (1980). *Upper Sacramento River spawning gravel study*. Red Bluff: California Department of Water Resources, Northern Division.

Potter, D. (1985). Sand sluicing from dams on the San Gabriel River—Is it feasible? In J. McGrath (Ed.), *California's Battered Coast, Proceedings from a Conference on Coastal Erosion* (pp. 251–206). California Coastal Commission.

Ramey, M. P., & Beck, S. M. (1990). Flushing flow evaluation: The north fork of the Feather River below Poe Dam. Environment, Health, and Safety Report 009.4-89.9. Pacific Gas and Electric Company, Department of Research and Development, San Ramon, California.
Reiser, D. W., Ramey, M. P., & Wesche, T. A. (1989). Flushing flows. In J. A. Gore & G. E. Petts (Eds.), *Alternatives in regulated river management* (pp. 91–135). Boca Raton, FL: CRC Press.
Richards, K. (1982). *Rivers: Form and process in alluvial channels* (p. 358). Methuen, London.
Schick, A. P., & Lekach, J. (1993). An evaluation of two ten-year sediment budgets, Nahal Yael, Israel. *Physical Geography, 14*(3), 225–238.
Schumm, S. A. (1977). *The fluvial system*. New York: Wiley.
Sen, S. P., & Srivastava, A. (1995). Flushing of sediment from small reservoir. In *Proceedings of Sixth International Symposium on River Sedimentation* (pp. 149–154). New Delhi, India.
Stevens, J. C. (1936). The silt problem. Paper No. 1927. Transactions American Society of Civil Engineers.
Stone, K. E., & Kaufman, B. S. (1985). Sand rights, a legal system to protect the shores of the beach. In J McGrath (Ed.), *California's Battered Coast, Proceedings from a Conference on Coastal Erosion* (pp. 280–297). California Coastal Commission.
Westrich, B., Al-Zoubi, S., & Muller, J. (1992). Planning and designing a flushing channel for river reservoir sediment management. In *5th International Symposium on River Sedimentation* (pp. 861–867). Karlsruhe, Germany.
Wilcock, P. R., Kondolf, G. M., Matthews, W. V., & Barta, A. F. (1996). Specification of sediment maintenance flows for a large gravel-bed river. *Water Resources Research, 32*(9), 2911–2921.
Williams, G. P., & Wolman, M. G. (1984). Downstream effects of dams on alluvial rivers. US Geological Survey Professional Paper 1286.

Chapter 6
The Legal Aspects of the Nile River

Abstract This chapter provides a brief overview of background of the legal status of the Nile water and the legal aspects of the Nile River; scope of the chapter focuses on the legal aspects of the Nile River, the Nile water treaties, treaties between U.K and the powers controlling the Nile Basin, the 1929 Nile waters agreement, agreements consolidating and supplementing the 1929 Agreement, state succession, state succession in the Nile Basin.

Keywords Scope of the chapter focuses on the legal aspects of the Nile River · The Nile water treaties · Treaties between U.K and the powers controlling the Nile Basin · The 1929 Nile waters agreement · Agreements consolidating and supplementing the 1929 Agreement · State succession · State succession in the Nile Basin · Succession to Treaties · The Problem in Perspective · The Claim that the Nile Water Treaties are valid and binding · The Case Against and historical significance: Cleaning up the Remnants of a Colonial-Legal Legacy?

The Nile is an international river. As a shared water resource, the development, utilization, and management of the Nile basin waters are regulated by international water resources law. Following the nomenclature of Article 38 of the Statute of the International Court of Justice, international water resources law may be derived from

(a) International conventions, whether general or particular
(b) International customs
(c) The general principles of law recognized by civilized nations
(d) As a subsidiary means, the judicial decisions and the teachings of the most highly qualified publicists of the various nations.

The conclusion of international treaties or conventions has been the most important method of international law-making, hence the primary means for the establishment of international rights and obligations over shared water resources. Most authorities would hold that an international treaty or convention is needed to ensure the most reasonable utilization of international water courses (Bruhacs 1993). There is no such international treaty applicable to the Nile, and even the

© Springer International Publishing AG 2017
Y.M. Hamada, *The Grand Ethiopian Renaissance Dam, its Impact on Egyptian Agriculture and the Potential for Alleviating Water Scarcity*,
Environment & Policy 55, DOI 10.1007/978-3-319-54439-7_6

United Nations Convention on the Law of the Non-Navigational uses of International Watercourses of 1997 (36 I.L.M. 700), which is sure to change the regime of international water law, has not entered into force.

Besides, although "nearly all the commentators on the problems of the full development of the Nile basin have concluded their various analyses with a suggestion in one form or another of the need for a Nile River Basin Authority or Administration" (Garretson 1960), such a basin-wide institution has never materialized.

The legal regime for the utilization and management of the Nile, therefore, consists of bilateral treaties concluded among the riparian states, and the international customary law. It has been suggested that these bilateral treaties reflect customary law principles (Fahmi 1986)—a position that has been vigorously contested (Batstone 1959; Pompe 1958).

Background

What is the legal status of the Nile water treaties described above—or more specifically, is the international legal regime established over the Nile through treaties concluded between Great Britain and Egypt with other powers still operational and binding on Nile basin states? The answer to this question is fundamental to the issue of rights and obligations over Nile waters.

If the Nile Waters Treaties are valid and binding, they legitimize the legal order of the colonial period that gave Egypt preeminence in the control of the Nile and developments in the basin. This would be a severe constraint on the development efforts and opportunities of upper riparian states. But if the Nile Waters treaties are not binding, then the control and utilization of Nile waters are regulated by the principles of customary international water law. It would also mean that the Nile is in search of a new legal regime in the form of a basin-wide agreement. This would provide plenty of room for negotiation and bargaining as among the riparian states. It could help develop a utilization regime that is more sustainable and equitable (Okoth-Owiro 2004).

Scope of the Chapter

This chapter of the book focuses on the legal aspects of the Nile River and the logic of a common East African position on the Nile question is dictated by a number of considerations. These include the pact signed by Egypt and Sudan in 1959, the fact that the East African countries are upper riparian, the idea of regional integration, their sharing of Lake Victoria, the history of sub-basin initiatives and the war in Sudan. Or the need to negotiate a legal and institutional framework for the management and utilization of the waters of the Nile has been canvassed in this case.

Such a "framework" should take the institutional form of a Basin Organization and the normative form of a restricted multilateral treaty. The states in the East African region need to take a common position on the Nile question and importantly to develop that position in preparation for negotiations with other riparian states.

The Nile Water Treaties

The Nile water treaties have been the subject of many studies and comments, most notably by Batstone (1959), Garretson (1960), Teclaff (1967), Okidi (1982, 1994), Godana (1985) and Carrol (1999).

As Godana (1985) observes, with the establishment of European colonial rule over most of the Nile basin in the closing decades of the nineteenth century, it became necessary to regulate, through treaties and other instruments, the water rights and obligations attaching to the various colonial territories within the basin.

In this manner, the colonial period came to witness a steady development of formal treaties and regulations as well as of informal working arrangements and administrative measures which, taken together, constituted the legal regime of the Nile drainage system.

The treaties and legal instruments regulating the use of Nile waters may be divided into four categories. These are as folllows:

(i) Treaties between the United Kingdom and the powers in control of the upper reaches of the Nile basin around the beginning of the twentieth century
(ii) The 1929 Nile Waters Agreement
(iii) Agreements and measures supplementing and consolidating the 1929 Agreement
(iv) Postcolonial treaties and other legal instruments.

Treaties Between U.K and the Powers Controlling the Nile Basin

Between 1891 and 1925, the United Kingdom of Great Britain entered into five agreements on the utilization of the waters of the Nile.

On April 15, 1891, the United Kingdom and Italy signed a protocol for the demarcation of their respective spheres of influence in Eastern Africa. Article III of this protocol sought to protect the Egyptian interest in the Nile waters contributed by the Atbara River, the upper reaches of which fell within the newly acquired Italian possession of Eritrea. The Article provided as follows: "The Government of Italy undertakes not to construct on the Atbara any irrigation or other works which might sensibly modify its flow into the Nile".

On May 15, 1902, the United Kingdom of Great Britain and Ethiopia, the former acting for Egypt and the Anglo-Egyptian Sudan, signed at Addis Ababa, a Treaty regarding the Frontiers between the Anglo-Egyptian Sudan, Ethiopia, and British Eritrea. Article III of the Treaty was concerned, not with boundaries, but with the Nile waters originating in Ethiopia. It provided: "His Majesty the Emperor Menelik II, King of kings of Ethiopia, engages himself towards the Government of His Britannic Majesty not to construct or allow to be constructed, any works across the Blue Nile, Lake Tsana or the Sobat, which would arrest the flow of their waters into the Nile except in agreement with his Britannic Majesty's Government and the Government of the Sudan".

On May 9, 1906, the United Kingdom and the Independent State of the Congo concluded a Treaty to Re-define Their Respective Spheres of Influence in Eastern and Central Africa. Article III of the Treaty provided: "The Government of the Independent State of Congo undertakes not to construct or allow to be constructed any work over or near the Semliki or Isango Rivers, which would diminish the volume of water entering Lake Albert, except in agreement with the Sudanese Government".

On April 3, 1906, the United Kingdom, France, and Italy signed a tripartite agreement and set of declarations in London. Article IV (a) provided that: "in order to preserve the integrity of Ethiopia and provide further that the parties would safeguard the interests of the United Kingdom and Egypt in the Nile basin, especially as regards the regulation of the water of that river and its tributaries …"

Finally, in December 1925, there was an exchange of notes between Italy and the United Kingdom by which Italy recognized the prior hydraulic rights of Egypt and the Sudan in the headwaters of the Blue Nile and White Nile rivers and their tributaries and engaged not to construct on the head waters any work which might sensibly modify their flow into the main river.

Garretson (1960) and Godana (1985) observe that regardless of whether the above agreements were concluded by Britain with another European power seeking to establish a sphere of influence, or with an African state such as Ethiopia, they had the common objective of securing recognition of the principle that no upper-basin state had the right to interfere with the flow of the Nile, in particular to the detriment of Egypt.

The 1929 Nile Waters Agreement

The Exchange of Notes between Great Britain (acting for Sudan and her East African dependencies) and Egypt in regard to the use of the waters of the Nile for irrigation purposes ("The 1929 Nile Waters Agreement") is the most controversial of all the Nile Water agreements. It is also the most important. According to Batstone (1959), it is the dominating feature of legal relationships concerning the distribution and utilization of the Nile waters today. Godana (1985) adds that the agreement "has become the basis of all subsequent water allocations (but) has been viewed differently by various writers" (page 176). The purpose of the 1929 Nile Waters agreement was to guarantee and facilitate an increase in the volume of water reaching Egypt. The Agreement was based on the outcome of political negotiations between Egypt and Great Britain in 1920s, and in particular on the report of the 1925 Nile Waters Commission, which was attached to the agreement as an integral part thereof.

The Agreement provided as follows

(i) Save with the previous agreement of the Egyptian Government, no irrigation or power works, or measures are to be constructed or taken on the River Nile or its branches, or on the lakes from which it flows in the Sudan or in countries under British administration, which would, in such a manner as to entail prejudice to the interests of Egypt, either reduce the quantities of water arriving in Egypt or modify the date of its arrival, or lower its level.

(ii) In case the Egyptian Government decides to construct in the Sudan any works on the river and its branches, or to take any measure with a view to increasing the water supply for the benefit of Egypt, they will agree beforehand with the local authorities on the measures to be taken to safeguard local interests. The construction, maintenance, and administration of the above-mentioned works shall be under the direct control of the Egyptian Government.

The Agreement also expressed recognition by Great Britain, of Egypt's "natural and historic rights in the waters of the Nile", even though the precise content of these rights was not elaborated.

The 1929 Nile Waters Agreement has been invoked by those who regard it as a praiseworthy recognition of the water rights of Egypt (Smith 1931). To some Egyptian writers, it has merely recorded Egypt's established rights over the Nile since antiquity (Khadduri 1972). But the overwhelming weight of expert opinion appears to favor the view that the "The 1929 settlement of the Nile" waters was a political matter and that it cannot be used as a precedent in international law (Berber 1959).

Agreements Consolidating and Supplementing the 1929 Agreement

The most important agreements falling into this category are the supplementary Agreement of 1932 (the Aswan Dam Project) and the Owen Falls Agreement. The last colonial-era treaty regulation of the Nile River System was the 1952 Agreement concluded by Exchange of Notes between Egypt and the United Kingdom (acting for Uganda) concerning the construction of the Owen Falls Dam in Uganda, then under British colonial administration. The purpose of the Agreement was twofold

(a) the control of the Nile Waters, and
(b) the production of hydroelectric power for Uganda.

The most important point of the substantive legal regime created by Owen Falls Dam Agreement was the regulation of the Nile River flow. The Agreement provided as follows: "The two governments have also agreed that though the construction of the dam will be the responsibility of the Uganda Electricity Board, the interests of Egypt will, during the period of construction, be represented at the site by the Egyptian resident engineer of suitable rank and his staff stationed there by the Royal Egyptian Government to whom all facilities will be given for the accomplishment of their duties. Furthermore, the two governments have agreed that although the dam when constructed will be administered and maintained by the Uganda Electricity Board, the latter will regulate the discharges to be passed through the dam on the instructions of the Egyptian Government for this purpose in accordance with arrangements to be agreed upon between the Egyptian Ministry of Public Works and the a pursuant to the provisions of agreement to be concluded between the two Governments."

The Agreement also provided that the Ugandan Government could take any action it considered desirable before or after the construction of the dam, provided that it did so after consultation and with the consent of the Egyptian Government, and provided further that: "—this action does not entail any prejudice to the interests of Egypt in accordance with the Nile Waters Agreement of 1929 and does not adversely affect the discharge of water to be passed through the dam in accordance with the arrangements to be agreed between the two Governments—."

In other words, the Egyptian interests in the flow of the Nile waters, as defined in the 1929 Nile Agreement, remained predominant, and Uganda's sovereign right to deal with its dam was made subject to the established and future Egyptian rights and interests. The Owen falls Dam was completed in 1954 (Okoth-Owiro 2004).

State Succession

State succession arises when there is a definitive replacement of one state by another in respect of sovereignty over a given territory in conformity with international law (Brownlie 1990). In other words, state succession consists of any change of sovereignty over a given territory whose effect is recognized in international law. It includes both "Succession in fact" and "Succession in law".

Succession in fact refers to the factual situation in which, through some political evolution, a territory that previously was placed under the sovereignty of one state comes to fall under that of another state, i.e., to the transfer of territory from one state to another. Such a transfer may occur when the territory of one state is annexed, in whole or in part; by another state, when one state cedes part of its territory to another; when two or more states merge to form a single state; when part of a national community secedes from a state, or combines with another existing state; or when a territorial community which was under colonial rule achieves independence by a process of revolution or constitutional evolution. The common feature of all these forms of factual succession is that one state ceases to be real in a territory and another takes its place.

Succession in law refers to the succession of the new sovereign to legal rights and obligations of the old sovereign, or more generally, to preexisting legal situations. Thus, succession in law is a legal consequence of succession in fact. We are here concerned with the obligations of the previous sovereign to its territorial successor.

State succession is an area of great uncertainty and controversy. This is due partly to the fact that much of the state practice is equivocal and could be explained on the basis of special agreement and various rules distinct from the category of state succession. Not many settled legal rules have emerged as yet (Brownlie 1990).

In other words, it is not clear, from either writings on international law or the practice of states, how and to what extent a legal principle of state succession applies in the sense of the transmissibility of rights and obligations from one state to another. For state succession in fact does not entail an automatic juridical substitution of the factual successor state in the complex sum of rights and obligations of the predecessor state (Godana 1985).

State Succession in the Nile Basin

The Nile Basin has witnessed several changes in territorial sovereignty over the years. Just as European occupation and colonization was the most important influence in state formation in the region, decolonization has been the most important cause of state succession. All the ten Nile Basin countries, except Egypt and Ethiopia, were dependencies of various European powers and became independent states in the second-half of the twentieth century. And aside from decolonization, state succession in the Nile basin has been prompted by such diverse factors as conquest, annexation, merger, and secession. A few examples will suffice.

Egypt has been part of the Ottoman Empire, under Turkish Suzerainty; a protectorate of Great Britain, an independent state and finally in 1958, part of the United Arab Republic after uniting with Syria. Ethiopia, a sovereign state, was conquered by Italy in 1936, a change of sovereignty that was recognized by European powers. Eritrea was a colony of Italy that became a part of Ethiopia and is now an independent state, after a war of secession. And Tanganyika, Rwanda, and Burundi were colonized by Germany and then became mandated territories, respectively, under Britain and Belgium, before gaining independence to become sovereign states. In 1964, Tanganyika merged with Zanzibar to form Tanzania. All the 10 riparian states on the Nile are successor states.

Succession to Treaties

The effect of change of sovereignty on treaties is not a manifestation of some general principles or rule of state succession, but rather a matter of treaty law and interpretation (O'Connell 1956). When a new state emerges, it is not bound by the treaties of the predecessor sovereign by virtue of a principle of state succession.

As a matter of general principle a new state, ex-hypothesis a non-party, cannot be bound by a treaty, and in addition other parties to a treaty are not bound to accept a new party, as it were, by operation of law (Brownlie 1990). The rule of non-transmissibility applies both to secession of newly independent states (that is, to cases of decolonization) and to other appearances of new states by the union or dissolution of states.

To the general rule of non-transmissibility (the "clean state" doctrine) there are some exceptions. The clear examples are

(i) law-making treaties or treaties evidencing rules of general international law
(ii) boundary treaties.

It is held by some writers that a third category of treaties, which they call "dispositive, localized" or "real," are also an exception to the general rule of no transmissibility (O'Connel 1956; McNair 1961). Proponents of the doctrine of dispositive treaties divide all treaties into two main categories, via, personal treaties and impersonal or dispositive treaties. Personal treaties are those dealing with political, administrative, or economic relations; they are, therefore, basically contractual in character in that they are personal to the parties. A personal treaty is said to be fundamentally a contract and, therefore, dependent on the continued existence of the parties. If any of the parties to such a treaty disappears in relation to a part of its territory, it ceases to be able to fulfill the obligations undertaken as a sovereign power over that territory.

Dispositive treaties, on the other hand, are those which create "real" rights and obligations, i.e., rights and obligations in rem in territory. As such, dispositive treaties are immune to the change of sovereignty and rem with the land like the easement of English Common law or the servitudes of Roman law. Examples of such treaties are said to include river treaties, boundary treaties, and treaties of peace and neutrality. The idea of dispositive treaties is unconvincing. Lester (1963) discusses it at length and finds first, that it is impossible to define the difference between localized and nonlocalized treaties, and second, that British state practice does not appear to recognize a special category of localized (dispositive) treaties for purposes of state succession. In his opinion, both in theory and according to British and Commonwealth practice, localized treaties are no exception to the general rule that bilateral treaties do not devolve upon successor states, and this opinion accords with the position in international law. Where rights in rem are recognized by new states, recognition is explained otherwise than an account of the automatic descent of treaties.

A similar conclusion is reached by Brownlie (1990) when he says; "The present writer, in company with others, considers that there is insufficient evidence in either principle or practice for the existence of this exception to the general rule. First, much of the practice is equivocal and may rest on acquiescence. Second, the category is very difficult to define and it is not clear why treaties apparently included should be treated in a special way". Supporters of the alleged exception lean on materials, which are commonly cited as evidence of an independent concept of state servitude (Brownlie 1990).

In practice, problems of succession are dealt with by devolution (or inheritance) agreements, by original accession to conventions by new states and by unilateral declarations (Brownlie 1990). On a considerable number of occasions, the inheritance or devolution of treaty rights and obligations has been the subject of agreements between the predecessor and successor states. Such agreements promote certainty and stability of relations. In Africa, Great Britain concluded inheritance agreement with Ghana, Nigeria, Sierra Leone, and the Gambia (Mutiti, 1976). It is, therefore, reasonable to conclude/infer that Great Britain did not want treaty rights and obligations over the Nile to devolve. Otherwise, she would have concluded inheritance treaties with her former dependencies.

The Problem in Perspective

The legal status of the Nile Water treaties has been a contentious issue since the 1950s. On May 18, 1956, in a statement attributed to the Joint Undersecretary for Foreign Affairs, it was stated that the British Government regarded the 1929 agreement and other treaties creating a regime over the Nile waters as subject to revision, and that it was intended to negotiate new terms on behalf of Kenya, Tanganyika, and Uganda (Lester 1963).

And August 27, 1959, the United Kingdom made the following statement. "— the territories of British East Africa will need for their development more water than they at present use and will wish their claims for more water to be recognized by other states concerned. Moreover, they will find it difficult to press ahead with their own development until they know what new works downstream states will require on the headwaters within British East African Territory". For this reason, the United Kingdom Government would welcome an early settlement of the whole Nile waters question (Garrestson 1960).

It is also a significant fact that as soon as the dependent territories became independent, they refused to accept the validity of the Nile Water Treaties. Thus, after attaining independence in 1956, Sudan denied the continued validity of the 1929 Nile Water Agreement. In fact, Egypt was compelled to negotiate a new treaty with its southern neighbor, the 1959 agreement on the full utilization of the Nile Waters. When it became independent in 1960, Tanganyika refused to be bound by treaties concluded by Great Britain on her behalf, and in particular, objected to the 1929 Nile Waters Agreement. In 1961, the Government of Tanganyika made a declaration to the Secretary-General of the United Nations in the following terms.

'As regards bilateral treaties validly concluded by the United Kingdom on behalf of the territory of Tanganyika, or validly applied or extended by the former to the territory of the latter, the Government of Tanganyika is willing to continue to apply within its territory on a basis of reciprocity, the terms of all such treaties for a period of two years from the date of independence—unless abrogated or modified earlier by mutual consent. At the expiry of that period, the Government of Tanganyika will regard such of these treaties which could not by the application of rules of customary international law be regarded as otherwise surviving, as having terminated' (Seaton and Maliti 1976; Brownlie 1990). Tanganyika's approach was adopted by other countries including Kenya, Uganda, Burundi, and Rwanda, who all refused to be bound by treaties concluded by colonial powers (Okidi 1982).

In the matter of the 1929 Nile Waters Agreement, the Government of Tanganyika, on July 4, 1962, addressed identical notes to the Governments of Britain, Egypt, and Sudan outlining the policy of Tanganyika on the use of the waters of the Nile. The note read as follows.

"The Government of Tanganyika, conscious of the vital importance of Lake Victoria and its catchment area to the future needs and interests of the people of Tanganyika, has given the most serious consideration to the situation that arises from the emergence of Tanganyika as an independent sovereign state in relation to the provisions of the Nile Waters Agreements on the use of the waters of the Nile entered into in 1929 by means of an exchange of Notes between the Governments of Egypt and the United Kingdom. As the result of such considerations, the Government of Tanganyika has come to the conclusion that the provisions of the 1929 Agreement purporting to apply to the countries under British Administration are not binding on Tanganyika. At the same time, however, and recognizing the importance of the waters of the Nile that have their source in Lake Victoria to the governments and people of all riparian states, the Government of Tanganyika is willing to enter into discussions with other interested governments at the appropriate time, with a view to formulating and agreeing on measures for the regulation and division of the waters in a manner that is just and equitable to all riparian states and the greatest benefit to all their peoples" (Seaton and Maliti 1976).

Another source of pressure on the legal status of the Nile Water Treaties is water stress and water scarcity in the Nile Basin. Hydrologists define countries whose annual water supply averages between 1000 and 2000 cubic meters per person as water stressed (the category before water scarce). A country is determined to be water scarce when its annual supply of internal renewable water falls below 1000 cubic meters waters per person (2740 L per day). In socioeconomic terms, scarcity occurs when the lack of water endangers food production, constrains economic development, and jeopardizes a country's natural systems (Gleick 1963). Due to a combination of factors, including population growth, consumption practices and patterns, diversionary activities of water resources and climatic and environmental conditions, the Nile basin countries are beginning to experience water scarcity, with four of them (Egypt, Kenya, Rwanda, and Burundi) already classified as water-scarce states (Kukk and Deese 1996). Access to the waters of the Nile is becoming a security matter, and the matter of rights and obligations is at the center of things. This pressure is compounded by the fact that most of the Nile Basin states have only recently started making systematic and appreciable (usually very unilateral) demands on the waters of the Nile and its effluents, as they embark on postcolonial programs of development. It is, therefore, not surprising that the status of the Nile treaties keep being raised as an issue in the flora in which resource rights and water are being discussed. In Kenya in recent times, there have been no less than four appeals to address the Nile waters question. Thus, speaking to journalists on February 12, 2002, Energy Minister Raila Odinga said that the 1929 Agreement should be renegotiated; Continued he; "The three countries (Kenya, Uganda and Tanzania) were not independent and were under colonial rule. That is what makes the treaty unfair. Why should we be denied the use of our water in the name of conserving it for others downstream?" (Daily Nation, 13th Feb, 2002, page 5).

And speaking at a water conference in Nairobi on March 21, 2002, a prominent international lawyer, Prof. Charles Odidi Okidi declared that the 1929 Agreement was not binding and should not be honored by Kenya and other East African countries (Daily Nation March 22, 2002, page 4). Mbaria (2002) and Kamau (2002) have made similar statements. The status of the Nile Water treaties has also been raised in the East African legislative Assembly by Yona Kanyomozi of Uganda (Kamau 2002).

The Claim that the Nile Water Treaties Are Valid and Binding

The claim (and assertion) that the Nile Water treaties are valid and binding on successor states is based on or encouraged by three sets of factors. These are the attitude of Egypt toward the treaties, the writings of certain publicists, and the ambivalent position expressed by some riparian countries.

(i) **Egypt's Position**

Egypt holds the view that all the Nile River agreements are by their nature perpetually binding on successor states. In her estimation, these instruments are transmitted to the successor states and may be either amended or abrogated only by consent in accordance with the Vienna Convention on the Law of Treaties. Egypt further asserts that treaties concluded by European powers acting on behalf of colonized African states continue to be in force by virtue of the law of state succession and because of the territorial nature of the obligations resulting from these treaties (Godana 1985).

Egypt also holds the view that she has "natural and historic" rights over Nile waters acquired by long usage and recognized by other states like Great Britain and Sudan, and that the Nile Water treaties have been declaratory of international customary law relating to fluvial law.

And from statements attributed to her political leaders, Egypt clearly regards access to the waters of the Nile as a national security matter. Egypt has repeatedly stated that if Ethiopia or any other upstream country diverts the Nile, she would use force to rectify the situation (Myers 1989; Starr 1991).

(ii) **The Writings of Publicists**

Some writers have expressed the view, based on the controversial idea of dispositive treaties, that the Nile Waters treaties were either declaratory of prescriptive rights or territorial in character and, therefore, transmissible. Thus Vali (1958) describes the 1929 Nile Waters Agreement as an agreement whose territorial character necessitates its respect by successor states (Lester 1963) and Godana (1985) proceeds on the assumption that the 1929 agreement is binding. He declares: "—Of all the early instruments on the utilization of the Nile Waters, only the 1929 agreement, as implemented by a number of subsequent agreements and measures, seems to survive. The survival of this particular treaty is unmistakably attested to by available evidence" (page 156).

But the "available evidence" is difficult to isolate, given that elsewhere, Godana (1985) opines as follows: "It seems doubtful that the 1929 agreement was seriously regarded or even intended as permanent in the sense that it would bind all successor states in perpetuity" (page 143).

Be that as it may, such opinions expressed by learned publicists create the impression, and encourage the interpretation that the Nile Water agreements are binding and valid, either because of their territorial character, or because it was the intention of the high contracting parties that the new sovereign states would be automatically bound by such treaties.

(iii) **The Uncertain Position of Some Riparian States**

Some Nile Riparian Countries have spoken strongly and consistently on the Nile Water Treaties, making it clear that they are not bound, and the treaties are not valid. These countries include Tanzania, Ethiopia, Sudan (on the 1929 Agreement), and Burundi. But there are states on the Nile basin whose positions have been rather ambiguous. A good example of such a state is Kenya. Even before independence, it was reported that "the local legislative councils of the territories (of East Africa, Kenya included) have indicated their dissatisfaction at what they consider to be the United Kingdom's inadequate international expression of their interests as upper riparian's" as regards the Nile Water treaties (Garretson 1960).

Then at independence, Kenya adopted the Nyerere doctrine and declared her intention not to be bound, giving a grace period of two years within which to renegotiate treaties entered into on her behalf by the United Kingdom. By a communication to the Secretary-General of the United Nations dated March 25, 1964, the Prime Minister of Kenya made the following declaration on the subject of succession of treaties extended or applied to Kenya by the Government of the United Kingdom prior to independence: "In so far as bilateral treaties concluded or extended by former Kingdom on behalf of the territory of Kenya or validly applied or extended by the former to the territory of the latter are concerned, the Government of Kenya is willing to be a successor to them subject to the following conditions:

(a) That such treaties shall continue in force for a period of two years from the date of Independence (i.e. until December 12, 1965).
(b) That such treaties shall be applied on a basis or reciprocity.
(c) That such treaties may be abrogated or modified by mutual consent of the other contracting party before December 12, 1965.

At the expiry of the aforementioned period of two years, the Government of Kenya will consider these treaties, which cannot be regarded as surviving according to the rules of customary international law as having terminated. The period of two years is intended to facilitate diplomatic negotiations, to enable the interested parties to reach satisfactory accord on the possibility of the continuance or modification or termination of the treaties (Mutiti 1976).

But recently, at a water conference in Nairobi, the Minister for Water Development, Mr. Kipngeno Arap Ngeny, inexplicably stated that the 1929 Nile Water Agreement was binding on Kenya" (Daily Nation, Saturday 23 March 2002, page 4). Some top government officials have even denied the existence of such treaties.

This kind of ambivalence encourages the assumption and belief that the Egyptian Government's position on the Nile is the true and legal position.

The Case Against

Finely this position is that the Nile Water agreements concluded during the colonial era are not binding on the successor states of the Nile basin and that this is the position in international law as buttressed by the practice of the states. The following reasons support this position.

First, the majority of commentators, with the distinct exception of Egyptians have come to the conclusion, or taken the position that these treaties are not binding, see (Godana 1985). The only controversial cases appear to be the 1929 Nile Water Treaty and the Owen Falls Agreement. Godana (1985) takes the view that the 1929 Agreement has "survived". O'Connell (1956) and others would take the view that such treaties are binding on successor states because of their "territorial character". However, the reasons adduced below make these treaties as invalid as any other colonial-era treaties.

Second, it is clear from the discussion in Chap. 4 that the strongest reason for claiming that the Nile Water treaties are binding is the doctrine of dispositive treaties. But it has already been shown that there is insufficient evidence for the existence of such a category of treaties as an exception to the general rule of non-transmissibility. Moreover, in the absence of the doctrine of dispositive treaties, some other basis for the survival of the Nile water agreements must be demonstrated. These alternative theories could be servitudes, acquiescence, or the idea of law-creating treaties (Lester 1963). But none of these have been shown to be the reason for the survival of the Nile treaties and authorities are in agreement that these theories are inapplicable to the case of the Nile water agreements.

Third, it has been implied by Egypt and some publicists that the validity of the Nile water agreements, their devolution on successor states, and their being binding in perpetuity is inferred from the intention of the parties. It is sometimes suggested that the description of a treaty as localized may refer to the intention of the high contracting parties with regard to the effect upon the treaty of alienation of territory to which it has been specifically applied, and that such intention might be "that the new sovereign will automatically be bound by the treaty" (Lester 1963). But the attitudes of Egypt and the United Kingdom and the provisions in the treaties do not evince such an intention.

Fourth, there is the doctrine of rebus sic stantibus. This doctrine asserts that if circumstances, which constituted an essential basis of the consent of the parties to be bound by a treaty, undergo such far-reaching changes as to transform radically the nature and scope of obligations still to be performed, the agreement may be terminated on the initiative of a party. It is submitted that the changes introduced by the decolonization process and the emergence of independent states in areas which were formerly territories under British administration are of such fundamental importance as to permit the operation of the doctrine of rebus Sistani bus, and that the declarations of the new states to the effect that the treaties entered into by former colonial powers on their behalf does not bind them is their initiative to terminate these treaties. Nile water agreements could, therefore, not survive colonialism.

Fifth, state practice is inconsistent with any claim of validity of the Nile water treaties. To this end, Great Britain had adopted the attitude that these treaties should be renegotiated and all states on the Nile basin (except Egypt) have adopted the Nyerere doctrine on state succession to treaties and have thus refused to be bound. Available evidence also shows that states on the Nile are taking unilateral decisions (or sub-basin approaches towards) in the utilization and development of Nile water resources.

Finally, and independently of the above, the conduct of Egypt with regard to the utilization of the Nile Waters raises serious doubts about her capacity to bind core patinas to their treaty and customary law obligations. (Chad 2009; Pompe 1958) submitted that the upper-basin states, already before their independence, "Could certainly not be held to the obligations which they undertook towards Great Britain by the aforementioned agreements of which Great Britain as administering power undertook towards Egypt by the 1929 agreement with regard to the construction of works affecting the Nile flow, if Egypt, or for that matter the Sudan were to construct dams which would change the natural conditions of the Nile Basin to the serious disadvantage of the upstream states" (Pompe 1958; Godana 1985).

In this connection, the building of the Owen Falls Dam (resulting in a rise of two and a half meters in the level of Lake Victoria), the Jong lei Canal project and particularly the diversion and piping of Nile water to Sinai Desert. (Okidi 1999; Mbaria 2002), and its reported sale to Israel would appear to be conduct that should release the upper riparian from any obligation towards Egypt. If Egypt can do as she pleases with the water, why should the other riparian be restricted?

The position adopted by Egypt on the legal status is dictated more by self-interest than by international law and state practice. That may explain her frequent resort to threats of military action and other "saber—rattling behavior."

Historical Significance: Cleaning up the Remnants of a Colonial-Legal Legacy?

Before European imperialists set foot in the Nile basin region, the Nile waters to Egypt were considered as a natural gift. Herodotus called Egypt "gift of the Nile," (Yohannes 2008) and Egypt and the Nile became intimately attached since then. Of course, Herodotus did not know the source of the Nile and how many other peoples depended on the Nile waters. Such a stravaganza, an Italian term for "extravagant" or "unreasonable", claim without proper knowledge of the issue is still fixed in the minds of many Egyptians today. This means also that the Nile Waters are viewed as a static, natural event that will be there forever. Egypt has, however, long been totally dependent on the Nile waters because of its arid climate and its resulting need to access fresh waters.

While Egypt's geographical challenges regarding access to fresh water resources have always been present, the arrival of European colonists into the region made the challenge even more complex. Especially, the arrival of colonial rivalry between Britain, France, and Italy was important. The British emerged strong from the colonial scramble for the Nile and it became British imperial policy to defend, control and manage the Nile waters (Collins 1990). For instance, Britain signed several treaties with various states of the Nile basin to protect its interests, particularly in Egypt, for the purpose of agricultural and cotton production (Madani et al. 2013), which was then shipped to England. It is from this backdrop that Egypt's historical and legal right was born, not out of the hydro-political interests of the basin-wide region.

Furthermore, Egypt and Britain strengthened their colonial-legal ties by signing the 1929 legal framework that would give Egypt veto power in all aspects of the Nile waters, including access and usage (Brunnee and Toope 2003). Colonial Britain signed the agreement not out of an altruistic agenda but as part of its own global imperialistic expansion, from which Egypt also benefited. **Terje Tvedt** confirms the benefits that Egypt received by stating that, "On 7 May 1929, **Lord Lioyd**, in a letter to the Egyptian Government, emphasized that Britain committed itself to guaranteeing Egypt its future water supply…" (Tvedt 2004) Expectedly, the colonial agreement between Egypt and Britain did not sit well for Sudan, which was at that point still regarded as part of Egypt and possibly to be crowned as a British colony (Tvedt 2004). In 1959, Sudan managed to generate a new legal agreement with Egypt after obtaining independence in 1956. The agreement was intended for the construction of the High Aswan Dam, which Egypt and Sudan had to ratify in order to build the dam (Allan 1999). The 1959 agreement between the two countries was also intended to renegotiate existing water allocations set out under the 1929 agreement (Cascao 2013). However, the agreement still did not include upstream riparian states (Ethiopia). **Cascao** affirms such development by stating that, "The 1959 agreement reinforced down streamer claims of 'natural and historical rights' to the Nile waters…and became the backbone of the hydro political dilemma in the Nile basin today".

References

Allan, J. A. (1999). The Nile basin: Evolving approaches to Nile waters management. Occasional Paper 20, SOAS, University of London. Accessed May 10, 2013. http://www.soas.ac.uk/water/publications/papers/file38363

Batstone, R. K. (1959). The utilization of the Nile waters. *I.C.L.Q, 8,* 523.

Berber, F. J. (1959). *Rivers in international law*. London: Stevens and Sons.

Brownlie, I. (1990). *Principles of public international law* (4th ed.). London: OUP.

Bruhacs, J. (1993). *The law of non-navigational uses of international watercourses*. Dordrecht: Martins Nijhoff Publishers.

Brunnee, J., & Toope, S. J. (2003). The Nile basin regime: A role for law? *Water Resources Perspectives: Evaluation, Management and Policy, 117*, 93–117. Accessed May 5, 2013. http://dx.doi.org/10.1016/S0167-5648(03)80010-2

Carrol, C. M. (1999). Past and future legal framework of the Nile river basin. *Geo-Int'l Envt'l L. Res, 12,* 269.

Cascao, A. E. (2013). Changing power relations in the Nile river basin: Unilateralism vs. cooperation? *Water Alternatives* (2009). Accessed May 28, 2013. http://agris.fao.org/agris

Chad, R. (2009). Writing British Columbia History, 1784–1958. UBC Press, The University of British Columbia, 2029 West Mall, Vancouver, BC v6t 1z2 FC3809.R445 2009. www.ubcpress.ca

Collins, R. O. (1990). *The waters of the Nile*. Princeton: Oxford University Press.

Fahmi, A. M. (1986). The legal regime of the river Nile. *OZOV, 37,* 51–70.

Garretson, A. H. (1960). The Nile river system. In *Proceedings of the American Society of International Law*. Washington D.C.

Gleick, P. H. (1963). Water in the 21st century. In P. H. Gleick (Ed.), *Water in crisis*. New York: CUP.

Godana, B. A. (1985). *Africa's shared water resources London*. Boulder Colorado: Lynne Rienner Publishers.

Kamau, J. (2002). *Can East Africa win the Nile War?* (p. 5). Daily Nation, March 28, 2002.

Khadduri, M. (1972). Other jurisdictional and territorial issues. In M. Khadduri (Ed.), *Major middle eastern problems in international law*. Washington, D.C: American Enterprises Institute of Public Policy Research.

Kukk, C. L., & Deese, D. A. (1996). At the water's edge: regional conflict and cooperation over fresh water. *UCLIA Journal of International Law and Foreign Affairs, 1,* 21–64.

Lester, A. P. (1963). State succession to treaties in the commonwealth. *I.C.L.Q, 12,* 475.

Madani, K., Rheinheimer, D., Elimam, L., & Connell-Buck, C. (2013). A game theory approach to understanding the Nile river basin conflict. http://www.tvrl.lth.se/fileadmin/tvrl/files/lb_day/97-114_Festskr.pdf. Accessed June 18, 2013.

Mbaria, J. (2002). *Revoke obsolete river Nile treaty* (p. 5). Daily Nation, March 28, 2002.

McNair, A. D. (1961). *The law of treaties*. Oxford: ONP.

Myers, N. (1989) Environment and Security. *Foreign Policy* 29.
Mutiti, N. A. B. (1976). *State succession to treaties in respect of newly independent African States*. Nairobi: E.A.L.B.
O'Connell, D. P. (1956). *The law of state succession*. Cambridge: CUP.
Okidi, C. O. (1982). Review of treaties on consumptive utilization of waters of Lake Victoria and Nile drainage system. *Natural Resources Journal, 22,* 161.
Okidi, C. O. (1994). History of the Nile and Lake Victoria basins through the treaties. In P. P. O. Howell, & J. A. Allan (Eds.), *The Nile: sharing a scarce resource*. Cambridge: CUP.
Okidi, C. O. (1999). Legal and policy considerations for regional cooperation on Lake Victoria and Nile River. *JEPLA, 1*.
Okoth-Owiro, A. (2004). The Nile treaty, state succession and international treaty commitments: A case study of the Nile water treaties. Konrad Adenauer Foundation, Law and Policy Research Foundation, Konrad Adenauer Stiftung and Law and Policy Research Foundation 2004.
Pompe, C. A. (1958). *The Nile waters question*. The Prague: Symbolae Verzijl.
Sahin, M. (1985). *Hydrology of the Nile river basin*. Amsterdam.
Seaton, E., & Maliti, S. T. (1976). *Tanzania treaty practice*.
Smith, H. A. (1931). *The economic uses of international rivers*. London: P.S. King and Sam.
Starr, J. R. (1991) Water Wars. *Foreign Policy* 17.
Teclaff, L. A. (1967). *The river basin in history and law*. The Hague: Mastinus Nijhoff.
Tvedt, T. (2004). The river Nile in the age of the British: political ecology and the quest for economic power. I.B. Tauris & Co Ltd.
Vali, F. A. (1958). *Servitudes of international law: A study of rights in foreign territory* (2nd ed.). London: Stevens.
Yohannes, O. (2008). *Water resources and inter-riparian relationships in the Nile basin: The search for an integrative discourse*. Albany: State University of New York Press.

Chapter 7
Agriculture and Irrigation

Abstract This chapter provides a brief overview of the history of agriculture and irrigation, background of agriculture and irrigation, scope of the chapter, water supply, improving performance and productivity of existing irrigation systems, improved basin and system-level management, riverine ecosystems impacts, on-farm technologies for enhancing the productivity of land and water, policy and institutional reform, improving alternative supply-side measures, enhancing rain-fed agriculture and supporting local techniques, adopting water recycling, investing in conventional supply-side measures and obstacles and enabling conditions.

Keywords Scope of the chapter · Water supply · Improving performance and productivity of existing irrigation systems · Improved basin and system level management · Riverine ecosystems impacts · On-farm technologies for enhancing the productivity of land and water · Policy and institutional reform · Improving alternative supply-side measures · Enhancing rain-fed agriculture and supporting local techniques · Adopting water recycling · Investing in conventional supply-side measures and obstacles and enabling conditions

Irrigation is the single largest consumptive use of fresh water in the world today. It is linked to food production and food security. About one-fifth of the world's agricultural land is irrigated, and irrigated agriculture accounts for about 40% of the world's agricultural production, yields from irrigated areas are on average double that of rain-fed agriculture, and are generally higher on land irrigated by ground water than on land irrigated with surface water (WCD 2000a, b). The total area irrigated expanded dramatically during the first years of the Green Revolution in the 1960s, increasing yields and bringing down food prices. From 1970 to 1982, global growth in the irrigated area slowed to 2% a year. In the post-Green Revolution period between 1982 and 1994, it declined to an annual average of 1.3% (Cosgrove and Rijsberman 1999).

Half of the world's large dams were built exclusively or primarily for irrigation, and an estimated 30–40% of the 268 million hectares of irrigated lands worldwide rely on dams. Discounting conjunctive use of ground water and surface water, dams are estimated to contribute to at most 12–16% of world food production. Ground

© Springer International Publishing AG 2017
Y.M. Hamada, *The Grand Ethiopian Renaissance Dam, its Impact on Egyptian Agriculture and the Potential for Alleviating Water Scarcity*,
Environment & Policy 55, DOI 10.1007/978-3-319-54439-7_7

water irrigation typically has higher yields, for a given amount of water, than surface water irrigation because of better control of the resource at farm level (Ibid).

Four countries—China, India, the United States, and Pakistan account for more than 50% of the world's total irrigated area. Dams supply the water for almost 100% of irrigated production in Egypt—most coming from the Aswan High Dam—while in Nepal and Bangladesh dams provide only 1% of irrigation water. In the two countries with the largest irrigated areas—India and China—official statistics suggest large dams supply approximately 30–35% of irrigation water, with the balance coming mainly from ground water sources. There is some controversy in calculating the percentage of food production attributable to dams, and particularly on the methods used to account for conjunctive use of surface and ground water, the WCD India Country Study noted that the official figures of the Central Water Commission indicate that 30% of irrigated land is supplied with water by dams; however, the study also suggested an alternative figure of 10% (WCD 2000a, b). Unsustainable irrigation practices have affected more than a fifth of the world's irrigated area in arid and semiarid regions. As a result, soil salinity and waterlogging either make agriculture impossible, or limit yields and the types of crop that can be grown. In other regions, overuse of tube wells has depleted ground water aquifers, lowering water tables, and making extraction increasingly expensive and especially difficult for smallholders. The absence of effective policies on conjunctive use of ground water and surface water resources is one of the most important concerns.

Background

Efforts to promote sustainable water management practices have necessarily focused on the agricultural sector as the largest consumer of freshwater. Governments have several objectives in deciding the nature and extent of inputs in agriculture. These include achieving food security, generating employment, alleviating poverty, and producing export crops to earn foreign exchange. Irrigation represents one of the inputs to enhance livelihoods and achieve economic objectives in the agricultural sector with subsequent effects for rural development.

Just as strategies and approaches to rural development are context specific, there are numerous and diverse alternatives to agricultural development and irrigation that need to be examined. The diversity relates to scale, level of technology, performance, and appropriateness to the local cultural and socioeconomic setting. Government policies and institutions play an important role in the promotion of particular water appropriation technologies and methods. Each method has different implications for food production, food security at local and national levels, and the distribution of costs and benefits.

The growth of modern 'conventional' irrigation since 1900 has been characterized by large water projects that harnessed rivers through the construction of diversion structures and canal systems. Since 1950, the spread of such technology accelerated through state-sponsored large-scale irrigation and an emphasis on large dams for water storage. Irrigated areas increased from 40 million hectares in 1900 to 100 million hectares by 1950 and to 271 million by 1998 (Postal 1999; FAOSTAT 1998). Dams support 30–40% of this area, with the remainder supplied from direct river abstraction, groundwater and traditional water harvesting systems (WCD 2000a, b). Since the 1970s, the predominant focus has been on providing irrigation to support the Green Revolution package of hybrid seeds, chemical fertilizers, and pesticides. Conditions for higher growth were created in such areas through subsidized infrastructure, agricultural inputs, and electricity for pumping.

Scope of the Chapter

Irrigated agriculture has contributed to growth in agricultural production worldwide, although inefficient use of water, inadequate maintenance of physical systems, and institutional and other problems have often led to poor performance. Emphasis on large-scale irrigation facilitated consolidation of land and brought prosperity for farmers with access to irrigation and markets. However, the scale of support to rain-fed areas was limited, even though such systems supported more than 80% of farmers in the developing countries of Asia and Africa. As a consequence, there has been a widening income gap between irrigated and rain-fed areas. Even within large-scale irrigation systems, inequities of this nature are observed, leading to the marginalization of small holders.

The regional economic and development context for agriculture differs markedly for industrial and developing countries. In the former, agriculture tends to be capital intensive with large, highly mechanized holdings requiring minimal labor. In contrast, agriculture in Asia and Africa supports hundreds of millions of smallholder cultivators who depend on land for subsistence, livelihood, and food security. These farmers generally do not have access to support mechanisms or capital resources to risk growing high-value crops in volatile market conditions. The low productivity of the land and labor of many subsistence cultivators is also symptomatic of absence of support and widespread neglect of their agriculture and irrigation systems (WCD 2000a, b).

Water Supply

Water supply dams in the WCD Knowledge Base have generally fallen short of intended timing and targets for bulk water delivery and have exhibited poor financial cost recovery and economic performance. These results reflect the longer development horizon of such dams, as well as overestimates of demand, and are similar to the general direction of results in the water supply and sanitation sector (WCD 2000a, b).

The Cross-Check Survey found that one-quarter of the 29 dams with a water supply function have delivered less than 50% of target. Furthermore, on average

70% of the sample did not reach their targets over time for the delivery of bulk water supply. Much of the overall sample variability is due to the multi-purpose, single-purpose distinction in conjunction with reservoir size. The data suggests that all delivery of bulk water in excess of planned targets can be ascribed to multi-purpose dams. This is contrary to performance trends for other purposes where single-purpose dams generally come out better. Regarding reservoir size, it is clear from the survey that the smaller the reservoir areas, the closer to target results have been—with the exception of the 11 reservoirs larger than 100 km^2. These very large reservoirs exhibit extreme variability, ranging from under-performance to considerable over-performance and delivering maximum achieved bulk water supplies up to 2.5 times planned targets (WCD 2000a, b). The principal message with regard to water supply emerging from the Case Studies is that even when it is not planned, demand for water supply from dams built for other purposes emerges over time. In the case of Tarbela, water from the river system downstream of the dam is diverted through irrigation canals to Karachi to supplement other sources of municipal water supply. In the case of the Aslantas, the growth of local districts has prompted an application to the authorities for the supply of over 400 million liters per day from the reservoir.

The Cross-Check sample showed a tendency toward underutilization of capacity that is reflected more generally in a 1994 post evaluation synthesis study of 31 water supply and sanitation projects by ADB. The utilization of capacity varied from 33–80%. Apart from the fact that evaluations were carried out at early stages in the life of the projects, a number of causes were suggested for low rates of utilization, including lower than expected growth in both population and in per capita consumption (ADB 1994).

Improving Performance and Productivity of Existing Irrigation Systems

There is considerable under-performance of large dam irrigation schemes and scope for improving the performance of existing systems. Increasing competition for water has highlighted the inefficiencies in irrigated agriculture and thus increased demands for a more effective and integrated approach to managing existing irrigation systems, particularly surface water irrigation. There is also considerable scope for enhancing the viability, adoption, and performance of other sources of irrigation water such as groundwater, direct river abstraction, and traditional water harvesting systems. In the past, emphasis on performance enhancement programs has had mixed results, however, with rapid decline in gains achieved leading to the need for periodic restoration, often heavily subsidized by the state. Policy interventions and institutional reforms to support technical interventions have been inadequate (WCD 2000a, b).

Improved Basin and System Level Management

The risk and consequence of sedimentation of reservoirs is site specific, but of considerable importance in a portion of the larger dam population. Measures to improve the sustainability of existing water resources systems through sediment flushing and catchment management can increase the contribution and longevity of irrigation systems. Enhancing infiltration and reducing surface erosion through catchment-protection initiatives may improve sustainability of reservoirs and irrigation systems, but must account for the trade-off with subsequent losses in annual water yield and the potential for lower dry season flows. A review of 94 catchment experiments from around the world suggests a loss in water yield of 10, 25, and 40 mm for a 10% increase in catchment cover, respectively, for scrub, deciduous hardwood and pine and eucalyptus (Bosch and Hewlett 1982).

Further, increases in vegetation lead to greater evaporation without necessarily leading to greater absorption and a reduction in surface run-off. Thus, the balance between evaporation and gains from increasing absorption will determine whether catchment measures lead to increases or decreases in dry season flow (Bruijnzeel 1990). The utility of these options will therefore be site specific and depend on the techniques applied and must be developed and evaluated in the larger context of natural resources management in the catchment. In particular the contrast between the effects on soil and water conservation of vegetative cover and structural measures such as embankments, ditches, and small dams needs to be considered.

Salinity affects approximately 20% of irrigated land worldwide (Postel 1999). Controlling salinity and reclaiming saline land is an urgent priority in order to increase productivity of existing land, make better use of irrigation, and demonstrate that new irrigation areas can be managed in a sustainable manner. Application of drainage technology and maintenance of existing drainage is one way of containing salinity. But reliance on physical drainage of saline effluent alone is insufficient to tackle the problem, and an integrated approach combining management of surface water, groundwater, and agricultural practices is essential. Salt-tolerant crops and vegetation can form part of such strategies to remove excess surface water and lower water tables. The irrigation of crops or pastures in "series' arranged in order of increasing salt tolerance is a further example of saline management. If practiced early enough, the integrated (or conjunctive) management of surface water, groundwater, and salinity can prevent the build-up of salts. Trials on this are currently under way in Australia and California.

Water quality is another important factor affecting basin productivity. For example, efforts to improve salinity problems upstream may have adverse effects downstream. An innovative approach to the problem of managing the disposal of saline effluent comes from the Murray Darling Basin of Australia, where disposal is regulated through regional load quotas or 'salt credits.' These credits are used to ensure highly saline water is released only during periods of high flow, when disposal has the least impact on river water quality. The irrigation system performance revealed a considerable lag time between commissioning of irrigation dams and full development of irrigation infrastructure. Accelerating the full development of associated infrastructure; such as canal networks and providing an integrated package of agricultural support measures for new irrigators may be a cost-effective option for enhancing performance.

In most irrigation systems, particularly those with long conveyance lengths, a disproportionate amount of water is lost as seepage in canals and never reaches the farmlands. For example, 40% of the water diverted from the Indus basin in Pakistan is lost in conveyance. In the late 1980s, it was estimated that improvements in supply efficiency could save some 14.8 billion m^3/year of water (WCD 2000a, b). Canal lining is one such improvement. Experience with buried plastic linings in China demonstrates good performance, over 18 years (Huasham et al. 1995). The United States introduced a low-cost lining program and since 1946 some 4600 km of various types of linings have been installed (Mitchell 1995). But in the absence of good quality control and effective maintenance, the canal linings often have not achieved the predicted improvements in water savings and reliability of supply. Studies from Pakistan in the early 1990s demonstrated that a focus on targeted maintenance is more cost effective than deferring problems to rehabilitation programs or lining canals (Murray-Rust and Vander Velde 1994).

Inadequate maintenance is a feature of a number of irrigation systems in developing countries. An impact evaluation of 21 irrigation projects by the World Bank concluded that a common source of poor performance was premature deterioration of water control structures (OED 1990). Often poor maintenance reduces irrigation potential and affects the performance of systems.

The scope for improvements through better maintenance is significant. The Food and Agriculture Organization (**FAO**) of the United Nations estimates that the total area that can be used with improved maintenance is 150 million hectares (FAO 1995). In India, such steps could add approximately 9 million hectares to the irrigated area (WCD 2000a, b). In countries like Nigeria, only 52% of irrigation from large-scale schemes was actually used in 1993 (FAO 1995). In the combined Gezira-Manage schemes in Sudan, 126,000 hectares had to be taken out of production due to sedimentation and weed growth in canals (FAO 1995). Performance-based maintenance contracts can be used as a cost-effective management measure to maintain system integrity. Such an arrangement can be extended to cover operation of the main canal system in tandem with decentralized farmer management at the secondary and tertiary levels. Proposals to improve system efficiency should be viewed in a river basin context examining the interactive effects of surface water, drainage reuse, and groundwater recharge (Seckler 1996).

Riverine Ecosystems Impacts

Growing threats to the ecological integrity of the world's watersheds come from rising populations, water pollution, deforestation, withdrawals of water for irrigation and municipal water supply, and the regulation of water flows resulting from the construction of large dams. Among the many factors leading to the degradation of watershed ecosystems, dams are the main physical threat, fragmenting and transforming aquatic and terrestrial ecosystems with a range of effects that vary in duration, scale, and degree of reversibility.

The watersheds of the world are the habitat of 40% of the world's fish species, and provide many ecosystem functions ranging from nutrient recycling and water purification to soil replenishment and flood control. At least 20% of the world's more than 9000 fresh water fish species have become extinct, threatened, or endangered in recent years (Revenga et al. 1998).

Fish are a critical source of animal protein for more than 1 billion people. In Africa, fish protein is 21% of the total animal protein in the diet, and in Asia it is 28%. While rivers supply about 6% of fish protein consumed by humans globally, it is often 100% of the supply for many inland riverine communities.

Ecosystem transformations do not only occur in the upper, lower, and mid-reaches of watersheds, but they also impact on river estuaries, which are frequently complex ecosystems. Closing the mouths of major rivers, salt intrusion, destruction of mangroves, and loss of wetlands are among the many issues at stake.

On-Farm Technologies for Enhancing the Productivity of Land and Water

Raising the efficiency of surface irrigation provides considerable scope for improvements within existing irrigation systems, whether the water comes from dams, rivers, or groundwater. There are large variations in efficiency. The average water use efficiency that was reported for the Atlanta's project by the **WCD** Case Study was 40%. However, the range of field-level efficiency for the project varied from 25 to 55%. Efficient surface irrigation, however, has produced figures of 60% (Cornish 1998). In a number of regions, irrigation has promoted cultivation of water-intensive crops such as sugarcane and rice. For example, intensive cultivation of sugarcane in India and Pakistan has led to unsustainable use of water. A number of technologies exist for improving water use efficiency and, hence, the productivity of water in irrigation systems.

Micro-irrigation methods, such as sprinkler and drip systems, provide an opportunity to obtain higher efficiency gains than those available in surface irrigation. Field application efficiencies are typically in the range of 70–90% (Cornish 1998). The output produced with a given amount of water is increased by allowing for more frequent and smaller irrigation inputs, improved uniformity of watering and reduced water losses. These methods have found wide usage in water-scarce regions. Nearly all of Israel's irrigated area, 68% of Jordan's, and 40% of Brazil's irrigated area uses micro-irrigation methods, as do small areas in China, India, and parts of Africa.

The capital-intensive nature of micro-irrigation technology and low cost of water has prevented more widespread adoption among smallholders in developing countries, but local manufacture has brought down the cost and improved viability for high-value crops. Further development in this area may make such pressurized irrigation technology attractive to farmers in developing countries. Low-technology versions, such as making portable drip line and bucket kits for small household irrigation, have recently gained support (Frausto 1999). Another method for increasing water use efficiency is to change crop cultivation practices. As with other water management initiatives, the potential for efficiency gains in water use needs to be examined within a river basin context, as in many cases losses upstream are recouped downstream.

Policy and Institutional Reform

Policy and management initiatives are fundamental to raising productivity per unit of land and water and increasing returns to labor. They are often interlinked and require political commitment and institutional co-ordination. Agricultural support program tends to be developed and implemented in relative isolation from irrigation systems. Typically there is weak co-ordination between agencies responsible for agricultural activities (such as extension services, land consolidation, credit and marketing) and those responsible for irrigation development. Price incentives are also inadequate to raise productivity and the outcome is a significant gap between potential and actual yields. In the absence of better opportunities from agriculture, many farmers seek off-farm employment.

Incentives to enhance production are necessary and can result from a more integrated set of agricultural support measures and the involvement of joint ventures that provide capital resources and market access to smallholder farmers. Appropriate arrangements need to be introduced for such joint ventures to ensure an equitable share of benefits.

The extent of recovery of the costs of operations and maintenance in large irrigation dams in the WCD Knowledge Base and irrigation systems in general is often limited. In a recent study of 16 projects, annual irrigation fees varied from zero in Thailand to as high as $130/ha in Colombia (FAO et al. 1999). Collection rates varied from 50 to 100%. One rationale for higher fees is that they will encourage more efficient irrigation practice and a shift to more water-efficient and higher value crops. For example, in the case of Pakistan and India reported above, farmers are attracted to the intensive cultivation of sugarcane by its profitability. This is because water charges are low, capital costs are not recovered, and the mitigation costs of attending to waterlogging or salinity problems are not borne by the farmer. Removing the subsidies inherent with supplying irrigation and drainage services may encourage adoption of technologies for increasing water use efficiency and promote a shift to less water demanding crops, especially in arid and water-scarce regions. Ideally, pricing structures for irrigation should reflect the cost of supplying water and associated externalities, and should be designed with stepped rates to provide security for basic livelihood needs.

In many systems, farmers are charged on a per-hectare basis independent of the volume of water used. This removes any incentive to save water. One obstacle to volumetric charges is the practical difficulty of measuring water delivered through an open canal system to a large number of smallholders. A solution may lie in water user organizations acting as intermediary bodies. They can enter into contracts with irrigation agencies and recover the charges directly from the irrigators. One of the major contributors to poor performance of large irrigation systems is the centralized and bureaucratic nature of system management, characterized by low levels of accountability, and lack of active user participation. Agency reform and management transfer have been initiated in more than 25 countries where governments are gradually reducing their roles in irrigation management and transferring responsibility for various levels of the systems to farmers' organizations and water user associations.

The major impetus for the transfer lies in the desire to cut back public expenditure on operation and maintenance costs (Vermillion 1997). The structure of farmer involvement varies from transfer of assets to a range of joint management models. As yet, there is no general evidence to suggest that irrigation performance has improved as a result of transfer alone, although there are promising examples indicating that decentralization may be a required, but not sufficient measure to improve performance (WCD 2000a, b; Vermillion 1997). Experience has shown that in order to be effective, a strong policy framework is required, providing clear powers and responsibilities for the farmers' organizations (Bandaragoda 1999; Vander Velde and Tirmizi 1999).

Water rights and trading are highly contentious issues. Win–win situations occur for farmers when they trade a part of their water to replace lost income while at the same time being able to finance water use efficiency gains from their remaining water allocation. In the United States, Colorado has one of the most advanced institutional support networks for water markets. In recent years some 30% of a district's annual water entitlements have moved through the rental market. The price at which farmers sell water is often significantly higher than their cost of supply. Enabling conditions for water markets are clear and secure entitlements along with effective administrative systems and infrastructure to regulate the trade and to monitor compliance. There should be clear environmental limits on the extent of the trade. These prerequisites are not present in many developing countries. Chile is often cited as an example where the development of water markets has helped to avoid expensive new water infrastructure, such as dams, by allowing transfers of water rights from agricultural to urban sectors (Brehm and Quiroz 1995; Hearne and Easter 1995; Hearne and Trava 1997). Concerns over water trading stem from the equity and livelihood implications of permanent trades of water, as opposed to yearly rentals, from small to large farmers or from rural to urban areas.

Improving Alternative Supply-Side Measures

Prior to the advent of large-scale irrigation, a number of traditional water appropriation techniques and irrigation methods were used in different contexts. Several supply options existed, complementing each other and contributing to agriculture, food production, and livelihoods. In today's context, local solutions and large-scale irrigation need not be mutually exclusive. They can co-exist and complement each other, raising food production and enhancing livelihoods.

For these systems to function optimally, a number of enabling conditions are required. Water appropriation systems need repair and maintenance, desalting, and weed clearing. Innovations and enhancement of traditional methods are needed to improve crop productivity. To optimize the productivity of the restored water management systems, appropriate land use, including cropping patterns, mix, and rotation need to be promoted. Sustaining these, location-specific systems and practices will depend on protection of sensitive catchments, floodplains, and deltas. Lastly, improving community stakes will be an important factor for greater application and long-term functioning (WCD 2000a, b).

Enhancing Rain-Fed Agriculture and Supporting Local Techniques

Some 80% of agricultural land worldwide is under rain-fed cultivation, contributing to 60% of food production (WCD 2000a, b). Given the number of low-income households that rely on rain-fed agriculture throughout the developing world, the enhancement of opportunities in this sector can have a major effect on productivity and livelihoods.

Over a period of time, these farming practices and irrigation methods have been marginalized by irrigation policies, the lack of institutional support and low levels of investment and research. In recent years, increasing attention has been paid to the successes of such methods and their importance is now being recognized by policymakers (Agrawal and Narain 1997; Thakkar 1999). The productivity of these rain-fed systems can be enhanced through improved agricultural support, local commitment to develop reliable water supply sources and small-scale, low-cost technologies. A progressive development path, where support provided is tailored to increasing local capacity, is emerging as an important conceptual approach to the development of such systems. For example, simple technologies such as treadle pump and low-cost drip systems can be introduced as a first stage innovation. In a few years—once farmers have recovered their investment costs—the returns can be reinvested in more advanced technologies such as small motorized groundwater pumps, thus gradually building the technological and capital base of the farmer (Frausto 1999).

Agriculture in rain-fed areas is supported by a number of water appropriation methods. Many of these have been used for centuries and have been adapted to satisfy local needs (WCD 1999). Rainwater harvesting uses small dams and embankments to trap run-off, usually with the aim of recharging groundwater.

Surface sources such as springs and streams are also diverted into natural depressions using bunds and check dams. In floodplains, cultivation uses natural moisture brought in by annual floods. A variety of water management techniques are used to tap the soil water in wetlands and valley bottoms. Many of these localized practices serve multiple purposes, such as checking saline ingress in coastal areas and recharging groundwater in floodplains. Effectiveness, sustainability, and productivity are location specific and depend on the extent to which they are spatially integrated. One successful example comes from the Rajasthan area of India. In many regions rainwater is harvested where it falls. In Argentina, Brazil, and Paraguay, natural or artificial depressions are used to store rainwater for crop and livestock production. In the semiarid areas of Argentina, Brazil, and Venezuela rainwater run-off captured from roads is collected in drainage ditches or street gutters and transported to cultivation areas (Ringler et al. 1999). Rainwater harvesting initiatives in Gansu province in China have provided both domestic and irrigation water to areas previously in drought. There are hundreds of thousands of small reservoirs or water tanks around the world, ranging from farm dams in Africa used to provide intermediate storage, to recession reservoirs in Cambodia that are used to enhance soil moisture for cultivation. In India and Sri Lanka, more than 500,000 tanks store rain water, sometimes supplemented by water from streams or small rivers (Agrawal and Narain 1997). Tank systems have provided irrigation and supported agriculture and livelihood for centuries in South Asia, although reliability problems, siltation, poor maintenance, and degradation have reduced the area served by small tanks in India from 4.6 million hectares in 1960 to 3.3 million hectares by 1987–88 (WCD 2000a, b). In addition to helping irrigation, tanks can play several important roles, such as flood control, soil erosion control, conservation of run-off during heavy rainfall, and recharge of groundwater.

The implications of the widespread replication of these small-scale storage methods on reducing demand for irrigation water and enhancing the sustainability of groundwater supplies, as well as their impact on existing surface water sources, could be far reaching. The floodplains and deltas of a number of the world's major rivers support wetland ecosystems of exceptional productivity that in turn support large rural communities. Floods provide natural irrigation that fertilizes floodplain soils (Acreman et al. 1999). As the floodwaters recede, arable crops are grown. Some soil moisture persists in the dry season and provides grazing for migrant herds. Floodplain cultivation is among the most productive and widely practiced of agricultural systems in central and sub-Saharan Africa. Similarly, wetland and delta cultivation is widely practiced in Southeast Asia. A form of flood agriculture specific to arid environments is spate irrigation, which diverts seasonal flood flows from dry riverbeds. Large areas under spate irrigation are found, for example, in Morocco, Yemen, and Pakistan (WCD 2000a, b). As noted earlier, there is a need to adequately account for the effects of structural alternatives such as dams and levees on these natural irrigation ‘options’. It is also necessary to consider the option of managed flood releases from existing facilities as a means of reestablishing and supporting these productive systems.

Adopting Water Recycling

Reuse of irrigation drainage water can be a major supply source in areas where intensive surface irrigation is currently practiced. Without formal arrangement, farmers in the Eastern Nile Delta in Egypt use approximately 3 billion cubic meters of saline drainage water every year for irrigation after suitable dilution or treatment. Similar examples exist in the North China plains, Arkansas valley in Colorado, the Pecos valley in New Mexico, and Australia (WCD 2000a, b).

Drainage water reuse requires increased levels of management skills to deal with increased levels of salinity, toxic concentrations, and associated health concerns. Reuse of urban wastewater is a significant source for irrigation in a number of countries.

In Israel, 275 million m^3 of wastewater are used for irrigation after treatment. This is approximately 22% of the total agricultural use of water (Shevah 1999). Strict controls are needed on the level of treatment required for various classes of water, with more stringent requirements for irrigating food grains than fodder crops. Examples from Ghana and Kenya also show the potential for peri-urban irrigation based on water reuse (DFID 2000). Where regulation of water quality is weak, this practice raises significant health concerns.

Investing in Conventional Supply-Side Measures

The potential to expand irrigation into new areas has sharply declined due to increasingly constrained resources and significant increases in the unit development cost. Conventional sources of water for irrigation besides reservoir storage behind large dams include diversion from rivers and lakes and groundwater abstraction. Diversion canals or lift irrigation pumping schemes supply irrigation systems by abstracting water from rivers. For example, river diversions and pumped irrigation serve 80% of the irrigated area in Kenya and 68% in Nigeria (FAO et al. 1999). The lack of over-season storage implies that the capacity of the system to provide multiple crops depends on the reliability of river flow. The size of run-of-river schemes can vary from a few hectares to hundreds of thousands. Run-of-river diversions can be used to supplement storage-based systems. In Sri Lanka, for example, diversion weirs were built to capture drainage water from upstream dam-based irrigation projects for reuse.

Groundwater abstraction has played an important role in the global expansion of irrigated agriculture. The availability of electricity, centrifugal pumps, and well drilling technology gave a major boost to rapid growth, particularly by individual farmers. Countries such as China, the United States, India, Pakistan, Bangladesh, Saudi Arabia, and the North Africa region have high rates of groundwater use. Currently in China, 8.8 million hectares of land are irrigated through groundwater wells, constituting 18% of the total irrigated area (PRC 1987). The Ogallala aquifer in the United States waters a fifth of that nation's irrigated land (National Research Council 1996). By the late 1990s, groundwater irrigated over half of all irrigated land in India, contributing to 78% of additional irrigated area created between 1984 and 1994 (Thakker 1999). Groundwater—employed on its own or in conjunction with surface irrigation—is often more productive than surface irrigation per unit applied. The determining factor appears to be the higher degree of control available to farmers who are often prepared to pay considerably more for reliable sources of supply such as groundwater (Dhawan 1998). Improved management of surface irrigation systems, leading to greater reliability of supply, may similarly increase production and returns to water.

Continued withdrawal of groundwater at current levels is, however, becoming unsustainable in many places. The Ogallala aquifer, for instance, is being depleted at the rate of 12 billion m^3 annually. Falling water increased pumping costs and historically low prices have led to a reduction in the area irrigated by the Ogallala of 20% over a 10-year period (WCD 2000a, b). Groundwater depletion has been a growing concern in the North China plains for over three decades, with water levels falling by 30 m since the 1960s (ADB 1999). Efficiency improvement measures are as important to groundwater systems as to surface water irrigation. Effective regulation is also necessary to curb over pumping, which leads to lowering water tables and can limit access of poorer farmers. Sustainable use of groundwater can be achieved through controlled abstraction and associated recharge measures. Recharge can be achieved by spreading surface water over large areas, using recharge wells, and water harvesting techniques as described above. Floodplains perform a significant natural recharge function. As floodwaters soaks through, underground reservoirs are recharged, and these supply water to wells beyond the floodplain. Recharge can also be a suitable approach for controlling saline intrusion and land subsidence and for reducing pumping costs. Most of the artificial recharge systems used to date have focused on small-scale systems or municipal water supply uses. Currently, almost half of the large dams in the world provide irrigation services. Irrigation projects can have significant multiplier effects on the local economy in terms of contributing to the development of agricultural processing and related industries. The **WCD** Knowledge Base does not elaborate on the comparative effects of the different options in terms of engendering such multiplier effects, although this will be an important consideration for options assessment (WCD 2000a, b).

Obstacles and Enabling Conditions

A number of policy, institutional, and regulatory factors hinder the emergence and widespread use of an appropriate mix of options that would respond to different development needs, sustain a viable agricultural sector, provide irrigation, and offer livelihood opportunities to large populations (WCD 2000a, b).

First, policy and institutional support for innovation, modernization, adaptation, maintenance, and extension of traditional irrigation and agricultural systems was lacking in the past. Increasing recognition of this has led a number of actors to place priority on improving rain-fed agriculture and developing small-scale irrigation capacity. An extension of this priority is the need to protect (or restore) the natural functioning of deltas, floodplains and catchments in order to sustain and enhance the productivity of traditional systems in these areas.

Second, the institutional framework needs to be redefined by transferring management to decentralized bodies, local governments, and community groups (water users associations or other appropriate bodies) for recovering tariffs and maintenance. Stronger commitment is required to transform irrigation bureaucracies into more efficient, service-oriented organizations capable of managing water and land in an integrated and sustainable manner. Irrigation water needs to be appropriately priced so that charges are based on volume used, taking into account the need to support basic needs and serve both equity and conservation.

Third, emphasis must be given to developing a package of agricultural support measures that are mutually reinforcing and develop intersectional linkages in the local economy so as to spur rural development. These efforts also need to counteract the tendency of current policies and extension services to impose high transaction costs and risks on smallholder farmers in developing countries. Measures to enhance security of tenure are also required.

Farmers also need access to international markets through reduction in barriers and supportive domestic policies. In response to structural adjustment programs and international agreements on world trade in agriculture, many developing countries have liberalized their agricultural policies, including cutting tariffs and subsidies. However, tariff and nontariff barriers to **OECD** markets—such as the large production and export subsidies for farmers in the US and European Union—limit the ability of developing countries' to diversify their agricultural sectors and gain the benefits of increased international trade, as well as increasing rural poverty where local farmers cannot compete with cheap imports (Smith 2000).

To sum up, future assessment of alternatives will need to clearly consider the following: improvements to the efficiency and productivity of existing irrigation systems before planning and implementing new ones; adaptation and expansion of local and traditional water management solutions; more co-ordinated management of surface and groundwater resources; and improvement of the productivity of rain-fed agriculture. To make progress in this area will require concerted efforts in policy formulation and institutional reform. As an aid to this assessment it will be useful to more fully understand the contribution of alternative irrigation and agricultural options to food production and livelihood security. For this purpose, analysis that extends beyond the immediate costs and benefits of these options to the secondary economic impacts on specific social groups is necessary (WCD 2000a, b).

References

Acreman, M. et al. (1999). Contributing paper for WCD, World Commission on Dams–2000, Thematic review II.1 Ecosystems.

ADB. (1994). Asian Development Bank Report, the 1994 annual report of the Board of Directors to the Board of Governors reviews ADB's operations, projects, internal administration, financial management.

Agrawal, A., Narain, S. (1997). Contributing paper for WCD, World Commission on Dams–2000, Thematic review IV.2 Irrigation options.

Bandaragoda, D. J. (1999). Institutional change and shared management of water resources in large canal systems: Results of an action.

Bosch, J. M., & Hewlett, J. D. (1982). A review of catchment experiments to determine the effect of vegetation changes on water yield and evapotranspiration.

Bruijnzeel, L. A. (1990). Hydrology of moist tropical forests and effects of conversion: A state of knowledge review. Humid tropics programmed, IHP-UNESCO.

Brehm, M. R., & Quiroz, J. (1995). The market for water rights in Chile: Major issues (Paperback). World Bank Publications, United States.

Cornish, K. (1998). The Jew of Linz is a controversial 1998 book.

Cosgrove, W. J., & Rijsberman, F. R. (1999). World Water Council change, social trends, and environmental quality

DFID. (2000). The Department for International Development (DFID) is the British government department responsible for promoting development.

Dhawan, R. (1998). Computes the interest cost of capital at book value (at historical prices) of the project at interest rates actually charged, rather than at opportunity; cited in WCD Thematic review IV.2 Irrigation options.

FAO et al. (1999). Collaboration with the FAO, land and water development division. Crop yield responses to deficit irrigation.

FAO. (1995). Reprinted by FAO, 1995. Published in 1995 by the Food and Agriculture Organization of the United Nations and Wiley, Baffin's Lane, Chichester.

FAOSTAT. (1998). Find your statistical data by exploring faostat data domains. ... of undernourished declining 1991–1998

Hearne, R. R.,& Trava, J. L. (1997). *Water markets in Mexico: Opportunities and constraints*. Environmental economics programmed discussion Paper, 97-01.

Hearne, R. R., Easter, K. W. (1995). Document search WTP315, Water allocation and water markets: An analysis of gains-from-trade in Chile.

Huasham et al. (1995). WCD thematic review IV.2 Irrigation options, Annex 8.

Liril HRISHITA THAKKAR. (1999). Contributing paper for WCD, World Commission on Dams–2000, Thematic Review IV.2 Irrigation Options.

Mitchell. (1995). WCD Thematic Review IV.2 Irrigation Options, p. 118.

Murray-Rust, D. H., Vander Velde, E. J. (1994). Report that farmers with access to a higher percentage share of canal water also pump more groundwater than farmers.

National Research Council. (1996). National Science Education Standards (1996). The standards are based largely on the 2011 National Research Council report A Framework for K-12 Science, cited in Postal, 1999.

OED. (1990). *The Concise Oxford English Dictionary is probably the best-known of the 'smaller' Oxford ... Eighth Edition (1990).* The Concise Oxford Dictionary of Current English, first edited by Fowler HW and Fowler FG was edited by Allen, R. E.

Postel. (1999). The decision of when and how much to irrigate are questions farmers have been considering for over 2000 years.

PRC. (1987). Ministry of water resources and electric power, cited in Postal, 1999, p. 56.

Rangel-Frausto, S. (1999). Contributing paper for WCD Thematic review IV.2 Irrigation options, p. 18.

Revenga, C., Murray, S., Abramovitz, J., & Hammond, A. (1998). *Watersheds of the world: Ecological value and vulnerability. World Resources Institute*. Washington, D.C. http://pubs.wri.org/pubs_description.cfm?PubID=2900.

Seckler, D. (1996). *The new era of water resources management, Research Report 1*. Colombo, Sri Lanka: International Irrigation Management Institute (IIMI).

Shevah. (1999). WCD Thematic review IV.2 Irrigation options, Section 4.3.2.

Scheuring, S., Ringler, P., Borgnia, M., Stahlberg, H., Müller, D. J., Agre, P., & Engel, A. (1999). High resolution AFM topography of the Escherichia coli water channel aquaporin Z. *The EMBO Journal, 18*, 4981–4987.

Smith. (2000). Contributing paper for WCD Thematic review IV.2, pp. 17, 30.

Thakker. (1999). Contributing paper for WCD Thematic review IV.2 Irrigation options, Frausto, op cit., WCD India Country.

Vander Velde, E. J., Tirmizi, J. (1999). *Irrigation policy reforms in Pakistan: Who's getting the process right?* Paper presented at International Researchers Conference on Participatory Irrigation Man-argument, Hyderabad, India, 11–14 December.

Vermillion, D. L. (1997). Impacts of irrigation management transfer: A review of the evidence, Research Report 11.

WCD. (1999). World Commission on Dams, India Country Study; Agrawal and Narain, op cit; Barrow, 1999.

WCD. (2000a). World Commission on Dams, India Country Study, Section 3.3.1.

WCD. (2000b). World Commission on Dams, Thematic review IV.2 Irrigation options. Yields from irrigated areas are on average double that of rain fed agriculture, and are generally higher on land irrigated by ground water than on land irrigated with surface water.

Chapter 8
Agriculture and Irrigation in Nile Basin

Abstract This chapter provides a brief overview of the history of agriculture and irrigation in the Nile basin, background of agriculture and irrigation in the Nile basin, scope of the chapter, the Nile river basin, scarce Nile water and existing sharing arrangements, move to establish basin-based cooperation, new water demands in the Nile basin, escalating conflict scenario, fear of climate change, The China factor and the riparian countries of the Nile understanding.

Keywords Scope of the chapter · The Nile river basin · Scarce Nile water and existing sharing arrangements · Move to establish basin-based cooperation · New water demands in the Nile basin · Escalating conflict scenario · Fear of climate change · The china factor and the riparian countries of the Nile understanding

For most of the twentieth century, the Nile River has been the source of political tensions and low-intensity conflicts among three of its major riparian countries (Ethiopia, Sudan, and Egypt). However, since the late 1990s, the Nile basin countries—with the encouragement and support of the international community—have made some attempts to establish basin-wide cooperative institutions. This process of engagement and collaboration is presently under severe stress due to increasing demand and decreasing supply of water resources in the basin. This situation may be complicated further by the global climate change, which is anticipated to result in long-term changes in the volume and pattern of run-off in the Nile River system. Moreover, the emergence of China as a major player in the power politics of the Nile basin has facilitated a number of unilateral initiatives for large-scale water development projects. In this context, this chapter critically examines the survival and sustainability of water cooperation endeavors in the Nile basin as the river faces challenges from the global climate change and shifting regional geopolitics (Swain 2011).

© Springer International Publishing AG 2017

Y.M. Hamada, *The Grand Ethiopian Renaissance Dam, its Impact on Egyptian Agriculture and the Potential for Alleviating Water Scarcity*,
Environment & Policy 55, DOI 10.1007/978-3-319-54439-7_8

Background

As far as the East is from the West, so far is, one may say, independent existence from colonial subjugation. This apparently incontrovertible truth though does not apply to the hydro-political and legal reality of the Nile basin as the postcolonial era is but an accentuated continuation, save for change of actors, of the ethos of the colonial era. During the twilight hours of British colonial rule in the basin, it was quite evident that Egypt, with its asymmetric power advantage vis-à-vis the other co-basin states and the imperial ambitions it has long had for complete control of the Nile water resources, precipitated by the ever-present fear of the possible interruption of the flow of the Nile by interference upstream, the passionate desire to gain control over the sources of the Nile with a view to ensuring the uninterrupted flow of the river downstream had for long been the major preoccupation of Egyptian rulers. The task was pioneered by Muhammad Ali (1769–1849) who drew a grand strategy of uniting the Nile Valley under Egyptian hegemony and unleashed a series of invasions which led to the conquest of Sudan in 1820. Muhammad Ali firmly believed that "the security and prosperity of Egypt could only be assured fully by extending conquests to those Ethiopian provinces from which Egypt received its great reserves of water" and used the conquest of Sudan as a stepping stone to launch repeated invasions along Ethiopia's western frontier; the campaign of conquest was brought to a halt when, in 1882, Egypt itself fell under British colonial rule. Kendie (1999), Brunnee and Toope (2002) would become the basin's bogeyman. When Sudan's independence was on the horizon, the campaign for uniting it with Egypt—a conviction birthed out of a traumatic experience which impressed upon Egypt's rulers "that whoever ruled Khartoum could hold Egypt for ransom" (Warbung 2000)—became a rallying slogan "viewed by Egyptian nationalists of all political shades as an absolute must" (Warbung 2000).

To the shock of Egypt and as a natural nationalist reaction, Sudan challenged, on the eve of its independence, the 1929 Agreement and called for its revision arguing that it "was no longer valid because it had been reached by Britain and Egypt [not involving Sudan] and it had discriminated against Sudan by granting it only one-twenty-second of the total annual flow of the Nile water" (Mekonnen 2011). It was quite understandable why the rallying slogan of the day—Unity of the Nile Valley—inspired, before Sudan's independence, in Egyptians and their Sudanese supporters "emotional and political significance similar to that inspired by 'Liberty, Equality and Fraternity' among the French revolutionists; 'The Union forever' among the Northern elements of the United States during the Civil War; the doctrine of 'Laissez-faire' among capitalists; or 'Workers of the World, Unite' among socialists" (Fabunmi 1960). Upon its independence in 1956, Sudan made it clear that the unfair terms of the 1929 Agreement would not bind it anymore and abrogated it 2 years later (Okidi 1980; Kliot 1958). The signing of a binding agreement on the utilization of the Nile waters by the two countries was made possible only "after the short-lived Sudanese parliamentary democracy was replaced by a military dictatorship led by General Abound" (Kliot 1958). The Agreement for the Full Utilization of the Nile Waters (Agreement 1959) provides nearly conclusive evidence of the fact that the hydro-political and legal reality in the Nile basin has not changed a bit, notwithstanding the demise of British colonial rule.

The central objective of the agreement was to realize the full utilization by the two parties of the Nile waters by replacing the 1929 agreement which "provided only for the partial use of the Nile waters and did not extend to include a complete control of the River waters" (Mekonnen 2011). The agreement made possible the launching of Nile Control Projects—the SudelAli (Aswan) and Roseires dams in Egypt and Sudan, respectively—which availed to the parties a net benefit of 22 billion cubic meters (Mekonnen 2011). The 22 billion cubic meter net benefit to be obtained from the Sud el Ali reservoir was then allocated to Egypt and Sudan which received further 7.5 and 14.5 billion cubic meters respectively on top of their respective historic rights reaffirmed by the agreement (Mekonnen 2011). The entire flow of the Nile was thus fully apportioned between Egypt and Sudan which received 55.5 and 18.5 billion cubic meters respectively, thereby entrenching a singularly iniquitous water utilization regime contingent upon zero consumptive water use by upstream riparian.

The history of Nile inter-riparian relationship has since been hallmarked with mutual distrust, aggressive unilateralism and open threats. Punctuated by occasional sabre-rattling in an atmosphere of intense bellicosity, the pattern of inter-riparian relationship has long been a tug of war between the two downstream riparian, which strive to endlessly perpetuate the status quo, and the upstream riparian, which call for its demise and replacement by an inclusive, fair, and equitable regime. This distinctively discordant pattern, entrenched for nearly half a century, began to change dramatically with the launch of the NBI in February 1999. The decade, since, has been a historic epoch of optimism and good rapport signifying, a remarkable shift in the tone and substance of state-to-state relationships along the Nile (Brunnee and Toope 2002). The NBI ushered in a fundamental transformation in the basin's history through an unprecedented inclusiveness in scope and an equally unprecedented depth in substance evident from the resolve to take up the sensitive issue of equitable reallocation which had consistently been eschewed by previous cooperative schemes (Mekonnen 2011).

Scope of the Chapter

The threat of water-related conflicts is comparatively more real and serious in the Middle East and North Africa hydrographic region where the Nile is found. Ominous predictions about water being the next casus belli in the region abound. There are many conflict determinants in the Nile basin which lend much credence to the predictions and the basin's proneness to conflict is quite evident. The unprecedented positive rapport brought about by the launching of the Nile Basin Initiative (**NBI**) and the enormous hope and optimism evoked by its lofty Shared Vision explain the unprecedented serenity and cooperative atmosphere the basin has witnessed over the past decade. The decade-long effort to work out and agree on an inclusive legal and institutional framework for the basin has, due to the cunning interpolation of the treacherous, nonlegal concept of 'water security', ended up in failure. The subsequent shift to and endorsement of benefit sharing as an alternative, simple and cure-all solution to the Nile waters question has further dimmed the prospect for the realization of the Shared Vision which now sounds more like a pipe dream than a realizable vision. Whether these adverse developments would finally pave the way for the ominous predictions to come to pass is as much unlikely as it is perplexing. It will be argued, in this chapter, that the likelihood of violent conflicts over the Nile waters is an unlikely scenario, the more likely turn of events being further continuation of the iniquitous status quo (Mekonnen 2011).

The Nile River Basin

The Nile Valley is referred to as "the cradle of civilization". However, in the post-Cold War period, due to increasing water demand along with the unequal distribution of water and power in the basin, many water researchers and policy makers have considered the same Nile River as having high potential to induce water war in the region (Clarke 1991; Gleick 1993; Swain 1996). Refuting all these dire predictions, and being actively encouraged and supported by the World Bank, the Nile Basin countries came together in 1999 to form the Nile Basin Initiative (**NBI**) with an objective to establish a basin-based water management institution. With this initiative, the Nile River is more often referred to in the water-related literature as an example of a shared river being the source of "water peace" rather than of "water war" in the basin. Figure 8.1 shows a map of the Eastern Nile and its sub-basins.

The Nile is probably the longest international river system in the world, with a length of 6670 km (Koutsoyiannis et al. 2008) (In 2007, researchers in Brazil claimed to have established a new source of the Amazon River, following an expedition to Peru and, if that claim is correct, the Amazon becomes the longest river at 6800 km and the Nile is relegated to second position.). The Nile River flows through 11 countries in Africa—Rwanda, Burundi, the Democratic Republic of Congo, Tanzania, Kenya, Uganda, Eritrea, Ethiopia, Sudan, South Sudan, and Egypt before reaching the Mediterranean Sea. Key statistics of the Nile River basin and of the riparian countries are presented in Fig. 8.1. The Nile receives its name from the Greek word "Neilos". The headwaters of the river are in Lake Victoria at about 4°S latitude, and then it flows mostly northward to the Mediterranean Sea at 32°N latitude. The Nile River has a drainage area of about 3.35×106 km^2, which covers 10% of the African continent. The White Nile and the Blue Nile are the two main tributaries of the Nile, which converge at the Sudanese capital, Khartoum, to form the main stream of the river. The White Nile originates in Burundi and flows through the Equatorial Lakes—Victoria, Kyoga, and Mobutu—and then passes through the Sudan swamps of Sudan (Swain 2011).

Fig. 8.1 The Eastern Nile and its sub-basins. *Source* Blackmore and Whittington (2008)

However, the Blue Nile, 1529 km long, which is the major water supplier, originates in the Ethiopian highlands. The Ethiopian highlands provide 86% of the Nile flow [comprised of Blue Nile: 59%; Baro-Akobo (Sobat): 14%; and Tekezze (Atbara): 13%], while the contribution from the Equatorial Lakes region is only 14% (Ethiopian Technical Experts 1996). The tributary Tekezze-Atbara has parts of its headwaters in Eritrea. The flow variations of the Ethiopian tributaries are very high between the flood and dry seasons. The Nile River is an extremely long water system and serves a large population in 11 countries; however, its average annual discharge is modest in comparison to the other major African river systems. While the average annual runoff of the Congo, Volta, Zambezi, and Niger rivers is 1250, 390, 230, and 180 km^3, respectively, that of the Nile River, whose catchment area covers 10% of the African continent, is less than 84 km^3 (Rangeley et al. 1994). The annual flow of the Nile is also moderate compared to the world's other major river systems: the Amazon River has annual discharge of 5518 km^3, the Mekong 470 km^3 and the Mississippi 562 km^3 (IUCN et al. 2003; UNEP et al. 2000).

As may be seen in Fig. 8.1, the total population of the Nile basin countries is 400×106, and over half of these people are dependent on the Nile for water supply. The Nile basin countries, particularly, Ethiopia, Sudan, and Egypt, have experienced massive population growth in recent years. If the present growth trend continues, the population of these three countries alone will swell to 340×106 by 2050. Owing to the growing population, the per capita water availability in the basin is decreasing rapidly. The Nile basin countries are among the poorest in the world. Apart from Egypt, Sudan, and Kenya, the Nile basin countries are among the least developed countries in the world. Surface water is very unevenly distributed in the region. The countries of the Great Lakes area are well endowed with water resources, while those downstream have scarce water supply. The Nubian Desert, through which the Nile passes to the Mediterranean Sea, is one of the driest regions on the Earth, and most of the population dependent on the Nile water inhabits this semiarid area. All the Nile basin countries have predominantly agricultural economies. To meet the needs of the growing population, irrigation has become essential for food security in the basin. Many countries in the basin are thus receiving a smaller share of the water resources; moreover, their agri-based economies need and consume more water (Swain 2011).

The agricultural sector is responsible for nearly 75% of total water withdrawal in the Nile basin. In both Egypt and Ethiopia, agriculture accounts for 86% of water withdrawal, while in the case of Sudan it is 94% (Mason 2004). Rainwater contributes significantly to the agriculture in the Great Lakes region countries. However, in the Lower Basin area, as can been seen in Table 8.1, the availability of rainwater is very scarce. In this part of the basin, even small-scale farming has to rely largely on river or groundwater. The two countries in the downstream area, Egypt and Sudan, are mostly dependent on irrigation. At present, out of a potential total area in

Table 8.1 Irrigation potential in different Nile sub-basins in Ethiopia

Nile sub-basin	Annual surface runoff (km^3)	Irrigated area in 1989 (ha)	Irrigation potential (ha)
Baro-Akobo	13.4	350	905,500
Blue Nile (Abbey)	54.7	21,010	1,001,500
Setit-Tekeze/Atbara	12.01	800	312,700
Total Nile basin	80.1	23,160	2,219,700

Source FAO (1997)

the Nile basin of 10.2 × 106 ha, approximately 5 × 106 ha are irrigated and 98.7% of this area is in the climatically dry Egypt and northern Sudan. It is important to point out that, while there are ten riparian countries, three of them are most critical for the meaningful cooperative sharing of the Nile water: Ethiopia as the primary supplier, and Egypt and Sudan as the dominant consumers (Swain 2011).

Scarce Nile Water and Existing Sharing Arrangements

The control of the colonial British administration over Egypt lasted from the late nineteenth century to 1937 and over Sudan from 1899 to 1956 (McCaffrey 1993). The other colonial powers having the presence in this region were Italy (over Eritrea and Ethiopia), France, and Belgium (over the countries of the Equatorial Region). However, Britain had always protected the Nile water flow to Egypt for strategic reasons. Several treaties were concluded among the colonial powers, giving priority to the Egyptian demands for the Nile water. However, after the construction of the first Aswan Dam in 1889, Egypt started to fear about the probable exploitation of the water resource in the upstream areas (Christiansen 1992). Egypt was increasingly becoming suspicious of British policy to increase cotton production in Sudan. Britain also started using the Nile water as a "carrot and stick" vis-à-vis Egyptian nationalism (Mohammed 1996). In 1929, Egypt and the British Government (on behalf of Sudan and the riparian countries of Lake Victoria) reached a water sharing agreement over the Nile water. Under the terms of the 1929 Agreement, Egypt was assigned the right to a minimum of 48 km^3 of water per year, while Sudan was assured to receive 4 km^3, leaving approximately 32 km^3 unallocated. However, this agreement did not include the major upstream water supplier, Ethiopia. The Agreement also noted that the East African countries were not to construct any water development projects in the Equatorial Lakes without consulting Egypt and Sudan. Egypt thus enjoyed the overwhelming rights, as against Sudan, in the utilization of the Nile water. The period of 1956–1958 witnessed serious disagreement between Sudan and Egypt over sharing of the Nile. Coinciding with the Sudanese objections to the proposed Aswan High Dam, Egypt withdrew from their previous agreement to enable Sudan to build the Roseires Dam on the Blue Nile. The relations deteriorated further when Sudan declared unilaterally its non-adherence to the 1929 Agreement (Warburg 1991).

After the military junta took over power in Sudan in 1958, being supported by Egypt, a new agreement on the sharing of the Nile River was reached in 1959 between Egypt and Sudan. From the newly calculated annual flow of 84 km^3 of water at Aswan, Egypt was to receive 55.5 and 18.5 km^3 were allotted to Sudan. The remaining 10 km^3 were allotted for mean annual evaporation and seepage losses from Lake Nasser behind the Aswan High Dam. The agreement also included some provisions for regulating the storage created by the Aswan Dam. Moreover, Egypt paid 15 million Egyptian pounds as compensation to Sudan for the resettlement of the 50,000 people displaced by Lake Nasser, which extended

150 km into the Sudanese territory. Sudan was granted permission to construct the Roseires Dam on the Blue Nile, as well as other projects necessary to utilize its own water share. Both parties also agreed to work on a cost-sharing basis to develop the water resources lost in the Sudanese swamps. Furthermore, both countries agreed that they would not negotiate with any third party over the Nile water before they had developed a common position. According to this agreement, the two countries signed a Protocol on January 17, 1960 to establish a Permanent Joint Technical Committee to facilitate technical cooperation on the projects (Okidi 1996).

However, Egypt did not have the financial resources to execute the High Aswan Dam project on its own. Meanwhile, Western powers were apprehensive about Egypt's closeness to the Soviet camp and their pressure forced the World Bank to withdraw from its commitment to support the High Aswan Dam project. In retaliation, the Egyptian President Nasser nationalized the Suez Canal and approached the Soviet Union for financial and technical support for the construction of the dam. With the Soviet support, the work on the High Aswan Dam started in 1960, and it came into operation in 1971. The Lake Nasser created by the High Aswan Dam is one of the largest manmade lakes in the world with the carrying capacity of 164 km^3 of water (Smith 1986).

The involvement of the Soviets in the dam building brought suspicion among the Western powers. With the help of the US Bureau of Reclamation, Ethiopia conducted a study to identify power and irrigation projects in its own Nile basin areas. The East African countries of the White Nile catchment areas, who were then under the British Administration, protested against their exclusion from the 1959 Agreement. However, in the post-Agreement period, the relationship between Egypt and Sudan became cooperative, primarily to maintain their domination over the Nile and not to allow other riparian countries to become successful in claiming their shares. The Islamic Sudan stood by Egypt even after the Camp David Accord and the Arab Summit in Baghdad. From the year of the Nile Agreement until the fall of the Sudanese President Numayri in 1985, Egypt had a friendly regime in Sudan, and received many concessions from Sudan. One of them was to carry out the Jonglei Canal project in 1976. The Jonglei I Project was planned to decrease the loss of water of the White Nile while it passes through the Sudan swamps in the south of Sudan (Collins 1990). The Jonglei I project was projected to have supplemented an annual flow of 4.7 km^3 of water, of which the share of Lake Nasser was 3.8 km^3. The construction of the project was started in 1978 by a French company. However, after 250 km of the proposed 360-km-long canal was completed, the work was forcibly suspended in 1984. A number of violent attacks on the construction site by the Southern Sudanese armed opposition, the Sudanese People's Liberation Army (**SPLA**), led to this decision (Suliman 1992). With this, the first serious effort to increase the yield of the Nile came to an end.

After Numara's departure from power in 1985, the relationship between Sudan and Egypt took a downward turn. As Sudan needs more water to meet the demand of its food production, there have been continuous demands from the Sudanese side to revise the 1959 Agreement in order to increase its share; however, Egypt was absolutely against it. In the 1990s, the Sudanese officials repeatedly started using the threats of withholding the Nile waters from Egypt. Sudan's alleged hand in the unsuccessful attempt on the life of the Egyptian President Hosni Mubarak at Addis Ababa in June 1995 brought further deterioration to the bilateral relationship. In this charged atmosphere, Hassan al-Turabi, the leader of the National Islamic Front (**NIF**) of Sudan threatened to stop the water to Egypt by redirecting the Nile's flow. This brought panic to the Egyptian authorities and initiated hectic official deliberations in Cairo. President Mubarak aggressively responded, "Those who play with fire in Khartoum …will push us to confrontation and to defend our rights and lives" (Sudan Update 1995). Similarly, his Foreign Minister declared, "I am warning Turabi not to play with fire, at the same time, not to play with water" (Nassar 1995). Egypt, in the past, had never hesitated to use the threat of war to stop the ambitions of the upstream countries in encroaching on the Nile's water. This tension between Egypt and Sudan was one of the main reasons for water analysts to fear "water war" in the basin in the 1990s (Clarke 1991; Gleick 1993; Homer-Dixon 1994; Swain 1996).

The Nile River has been the cornerstone of the Egyptian regional foreign policy. Egypt always proclaims its historic right to utilize the waters of the Nile, which goes back to 5500 years. The other riparian countries have alternative sources of water supply, while Egypt is almost completely dependent upon the Nile for its water supply and irrigation. Thus, Egypt is very sensitive about water diversions in the upstream regions of the river. However, the question being increasingly posed is whether Egypt can afford to continue to use large quantities of water for agriculture and other purposes when the needs of other countries in the upstream areas are also growing. Ethiopia now plans to harness the waters of the Blue Nile, and Kenya, and Uganda wish to develop the water resources of Lake Victoria. Since the late 1980s, the situation has remained tense among the three major riparian countries of the Nile: Egypt, Sudan, and Ethiopia, with Egypt regularly threatening to use its military muscle to protect its share of the Nile water (Dawoud 2001).

However, the major threat to Egypt's water supply came from Ethiopia in the mid-1990s. With its own rapid population growth and increasing food demands, Ethiopia needed more water for its own use. Not constrained by any water-sharing agreement with Egypt or Sudan, it unilaterally developed plans to divert the Nile waters for its own irrigation projects. In spite of objections from Egypt and Sudan, Ethiopia maintained its sovereign right to develop the water resources within its borders. Economic and technological backwardness and political troubles had stalled the Ethiopian plans to develop Nile water for a long time. In the mid-1990s, Ethiopia emerged from a long period of civil war and famine into a period of accelerated growth and economic development. The financial assistance from the Western countries increased considerably in the 1990s as well. This provided confidence in Addis Ababa to successfully counter the Egyptian and Sudanese objections to its water development projects at the diplomatic level. Since the mid-1990s, the Ethiopian government has constructed a large number of small dams, and many additional dams are also planned, to enhance the country's irrigation and hydropower capacity. Ethiopia's current water development plans require a small portion of the Nile's water compared to the potential demand in the near future. It intends to develop its hydropower potential in order to meet the increasing energy demands.

However, the real threat to downstream water supply does not come from hydropower production, which does not divert water from the river system, the way irrigation does. Ethiopia has an estimated 2.2×106 ha of land to be irrigated and its rapid population growth and economic liberalization demand the development of a large portion of this land for agricultural purposes, which could significantly threaten water supply of the lower Nile riparian countries (Swain 2011).

Move to Establish Basin-Based Cooperation

Since the 1960s, there have been some unsuccessful attempts to achieve a basin-based arrangement on the Nile River system. In 1967, Egypt, Kenya, Sudan, Tanzania, and Uganda, with the United Nations Development Programmed (**UNDP**) and the World Meteorological Organization (**WMO**), came together to launch the Hydrometric Project. The purpose of the project was to evaluate the water balance of the Lake Victoria catchment in order to assist in any control and regulation of the Lake level as well as the flow of water down the Nile. Later, Rwanda and Burundi joined in the effort. This Project lasted for 25 years, but did not include the major water contributor Ethiopia, and failed in its effort to develop an effective basin-based arrangement. The Hydrometric Project was completed in 1992 and, in the same year, a new body, the Technical Committee for the Promotion of the Development and Environmental Protection of the Nile Basin (**TECCONILE**), was created. Water resource ministers from Egypt, Sudan, South Sudan, Rwanda, Tanzania, Uganda, and Zaire (later renamed as Democratic Republic of Congo) signed an agreement to this effect in Kampala in December 1992. The other four riparian states participated as observers. The **TECCONILE** came into operation on January 1, 1993, with its Secretariat at Entebbe, Uganda. An important parallel activity in the effort to bring about basin-based cooperation was the Nile 2002 Conference series, which started in 1993 and continued up to 2002. Each year, these conferences assembled technical experts from all the Nile basin countries with the aim of providing a forum for scientific discussion and informal dialogues to exchange views and foster cooperation for Nile basin development.

In March 1998, the Council of Ministers of Water Affairs of the Nile Basin States reached a broad agreement at Arusha, Tanzania over the sharing and managing the Nile water. The Arusha meeting was attended by eight out of 10 riparian states. The meeting, in the absence of participation by Eritrea and the Democratic Republic of Congo, endorsed a new programmed of action. This led to the formal launching of the Nile Basin Initiative (**NBI**) in February 1999.

Member countries are: Burundi, Democratic Republic of Congo, Egypt, Ethiopia, Kenya, Rwanda, Sudan, South Sudan, Tanzania, and Uganda; Eritrea holds observer status. In September 1999, the NBI Secretariat was officially opened in Entebbe, Uganda. The NBI is a transitional arrangement until the member countries agree on a permanent legal and institutional framework for sustainable development of the Nile basin. The **NBI** supersedes the disbanded **TECCONILE** (Swain 2002).

The Nile basin countries have formally agreed on a "Shared Vision" that seeks "to achieve sustainable socio-economic development through the equitable utilization of, and benefit from, the common Nile basin water resources." The objectives of the **NBI** are

1. to develop the water resources of the Nile basin in a sustainable and equitable way to ensure prosperity, security, and peace for all its peoples;
2. to ensure efficient water management and optimal use of the resources;
3. to ensure cooperation and joint action among the riparian countries, seeking win–win gains;
4. to target poverty eradication and promote economic integration; and
5. to ensure that the programmed results in a move from planning to action.

To achieve these objectives, the countries have devised a Strategic Action Programmed (**SAP**) that promotes a Shared Vision Programmed (**SVP**) and two investment Subsidiary Action Programs. The **SVP** is designed to build the institutional capacity, relationships, and technical skills needed to support the NBI's two investment programs: the Nile Equatorial Lakes Subsidiary Action Programmed (**NEL-SAP**) and the Eastern Nile Subsidiary Action Programmed (**ENSAP**).

The Strategic Action Programmed is meant to translate the shared vision into concrete activities through a twofold, complementary approach

- To lay the groundwork for cooperative action through a regional programmed to build confidence and capacity throughout the basin (the Shared Vision Programmed).
- To pursue, simultaneously, cooperative development opportunities to realize physical investments and tangible results through sub-basin activities (Subsidiary Action Programs) in the Eastern Nile and the Nile Equatorial Lakes regions.

The joint development of the Nile waters requires significant financial resources. The World Bank coordinated the launch on January 26, 2001 at Geneva, Switzerland, of an International Consortium for Co-operation on the Nile (**ICCON**), where donors pledged to raise an initial amount of US$140 million, and expressed strong support to finance the first phase of the investment programmed, which was expected to be US$3 billion. The then President of the World Bank, **James Wolfensohn**, had taken personal interest in the Nile Basin Initiative (Wolfensohn 2001). In 2003, at the request of the Nile Basin Council of Ministers, the Nile Basin Trust Fund (**NBTF**) was established. The NBTF helps to administer and harmonize donor partner support pledged to the Nile Basin Initiative. Canada, Denmark, The Netherlands, Norway, Sweden, and the United Kingdom are some of the donors who contribute through the **NBTF**. Thanks to pressure from the World Bank, Egypt has agreed to bring about a superficial shift in its regional foreign policy over the Nile water issue. The adverse economic situation and the need for financial support by the World Bank practically forced Egypt's decision to express its willingness to cooperate, and not to explicitly demand a monopoly, over the Nile water. However, expressing concern and taking concrete actions are two different things. Historically, Egypt has prevented upstream development in order to maintain its control of the Nile. There has been no recent reduction of its dependence on the Nile water. Rather, Egypt's demand for water is increasing considerably. It needs vast quantities of water to increase agricultural production to feed its growing population. With the Nile almost being the only source of its water supply, Egypt has a limitation to concede to the Ethiopian demands for water development and is continuing to plan and carry out large new water projects unilaterally within its borders.

New Water Demands in the Nile Basin

In the early 1990s, the Nile River was the direct source of more than 55 km^3 of Egypt's total water use of 65 km^3, while the rest mostly came from the underground sources and water reuse and treatment (Hvidt 1995). Owing to the increasing population, the need for water in Egypt is also rising (Postel 1995; Mason 2004). Egypt's water demand has increased further due to greater irrigation works resulting from land reclamation projects. In 2000, the total water demand of Egypt increased up to 73.3 km^3, of which the share of the agricultural sector was 60.7 km^3.

The agricultural sector water demand is projected to reach 69.43 km^3 by 2025, out of a projected total water demand of 86.89 km^3 (Farrag 2005). Egypt is still heavily dependent on imported food, with nearly 50% of the food coming from abroad (Sehmi 1996). To achieve some sort of food security in the face of rapid population increase, Egypt needs more and more water, particularly for irrigation use, from the Nile River (Swain 2008). Due to this, as Hassan and Rasheedy (2007) argue, "the Nile River constitutes the central environmental factor, which influences the cognitive behavior of the Egyptian foreign policymakers". Though Egypt has already been using most of the Nile water, it plans to use even more. Increasing demand for food in the future will certainly bring further pressure on the scarce water supply. Egypt also plans to create new urban centers and industrial areas in the desert to make living space for over one-fifth of its population. Moreover, Egypt's water requirements have further increased due to greater irrigation works resulting from land reclamation projects in the middle of its Western Desert (El Quosy 2001). The Southern Valley Agricultural Development Project (the Toshka Project) in the Western Desert is undertaken in order to resettle millions of people and provide irrigation to more than 200,000 ha of desert by transporting 5.5 km^3 of water from Lake Nasser through a 310-km-long pipeline (Gardner-Outlaw and Engelman 1997). This land reclamation project in the Western Desert has even brought tension within Egypt, as it is feared that the water supply to other areas would be affected. Moreover, all these ambitious projects have added to the economic troubles of the country.

Egypt has built several irrigation projects, such as the Isna Barrage, the Nag Hammadi Barrage, the Asyut Barrage, the Damietta and Rosetta deltas, the Zifta Barrage, the Idfina Barrage and the Damietta Dam in the downstream (El-Moattassem et al. 1996), which will be certainly affected by the reduced flow from Aswan. The other water-guzzling project is the Sinai Development Project, which increases the amount of land that may be irrigated by mixing water from the Nile with drainage water. In 1997, President Hosni Mubarek of Egypt opened the 261-km-long Al Salaam (Peace) Canal to transport water under the Suez Canal into the Sinai. Egypt is now busy constructing new projects on the Nile and bringing new desert areas under cultivation, thus making it further dependent on the Nile water. Furthermore, the upstream countries of the Nile are making serious efforts to use the Nile water for their own needs. Adding to Egypt's precarious water situation, the evaporation from the surface of the 600-km-long Nasser Lake apparently exceeds the earlier estimated amount.

Sudan is the largest country in the African continent, connecting the Arab World to Sub-Saharan Africa. Its population of 42 × 106 is growing at an annual rate of over 2%. While the country's average annual rainfall is 416 mm, the share of the dry north is only 25 mm compared to over 1600 mm in the rainforest areas of the south. Thus, the northern part of the country, which receives virtually no rainfall, is dependent completely on irrigation for agriculture. While in Egypt 50% of the population is dependent on agriculture, in Sudan the number is more than 70%. This prompts some to argue that Sudan is more dependent on the Nile than Egypt is Sofer (1999).

The high seasonal fluctuation in the river runoff has prompted Sudan to build dams on the Nile system to store the water of the Nile for flood control and irrigation purposes. Before the construction of the High Aswan Dam, the Jebel Aulia Dam was built on the White Nile to support the dry season flow to the Aswan Dam. After the 1959 Agreement with Egypt over the sharing of the Nile water, few new dams were built in Sudan. The prominent ones are the Roseires Dam and Sennar Dam on the Blue Nile and the Kashm el Girba Dam on the Atbara River. Sudan has also constructed a number of smaller barrages in Blue Nile system. Due to internal conflict, the country's plan to build the Jonglei Canal, to divert water from upstream of the Sudan to a point farther down the White Nile, bypassing the swamps, to make more water available for use downstream has not been realized. At the same time, the storage capacity of Sudan's four main reservoirs on the Nile system has been reduced due to increasing sedimentation.

Sudan has managed to cultivate only 16.7 × 106 ha of its land out of a potential 105 × 106 ha. Moreover, it has managed to provide irrigation to only 1.9 × 106 ha out of a potential 2.8 × 106. Sudan is in need of more water to meet the demand of its food production. Its plan to bring more agricultural land under the scope of irrigation creates a further demand for water. The increased desertification and land degradation have multiplied the country's water problem. Sudan claims that it has virtually consumed its share of the Nile water. The potential demand of Sudan for the Nile water is estimated to increase to 32 km^3 by the year 2025, particularly due to increased irrigational use. Sudan has planned to construct a number of new dams on the Nile (The Republic of Sudan 1996). It is also true that, until recently, Sudan had only a limited capacity to develop large-scale water projects unilaterally. However, Sudan has received increasing geopolitical importance in recent years. The growing direct foreign investment from many Arab countries and also from China, and its own massive oil revenue, has transformed the country to become a real competitor for acquiring more Nile water resources.

Sudan is presently constructing a number of large hydropower projects in the Nile. Furthermore, it has concrete plans to extend its capacity of irrigated agriculture and that will take its water abstraction from the Nile above its quota stipulated in the 1959 Agreement. Gradually, Sudan is moving toward a serious challenge to Egyptian domination on the basin. Besides Egypt and Sudan, Ethiopia carries significant importance in the sharing and management of the Nile River. The Ethiopian highland region contributes nearly 86% of the Nile flow, which rises to 95% during the flood period. This massive upstream contribution confers an undoubtedly key position to Ethiopia on the management of the Nile water. Moreover, Ethiopia is not bound by any agreement with Egypt and Sudan over the sharing of the river.

Agriculture accounts for 40% of the Ethiopia's gross national product (**GNP**), 90% of the export income and provides employment to 85% of the population. Ethiopia's large agricultural population is located in high-rainfall, highland areas. Ethiopia presently cultivates 90,000 ha of irrigated land, which is only 4% of the potentially irrigable land in the country. As can be seen in Table 8.1, Ethiopia has more than 2.2×106 ha of irrigable land in its part of the Nile basin, of which approximately 1% has been developed. Ethiopia regularly experiences frequent and large shortfalls in food production, leading to increased food imports. Ethiopia has witnessed two severe famines in the last 25 years, which have also led to civil unrest and subsequent fall of the regimes. The present regime in Ethiopia is serious about achieving self-sufficiency in food production at any cost (Swain 1997, 2008). For their own political future, Ethiopians do not want the reoccurrence of the famines of 1970s and 1980s on their land.

Ethiopia's irrigation plans could substantially reduce the water flow in the Nile. Given her need to increase food production, and with the headwaters of the Nile being one of its few natural resources, Ethiopia intends to use more of the Nile water for her own use. As Ethiopia's Prime Minister, Meles Zenawi, asserted (The Egyptian Gazette, 20 May 2010): "Some people in Egypt have old-fashioned ideas based on the assumption that the Nile water belongs to Egypt, and that Egypt has a right to decide who gets what, and that the upper countries are unable to use the Nile water because they will be unstable and they will be poor. These circumstances have changed and changed forever. Ethiopia is not unstable. Ethiopia is still poor, but it is able to cover the necessary resources to build whatever infrastructure and dams it wants on the Nile water."

Ethiopia maintains her sovereign right to develop the water resources within its borders. Economic and technological difficulties and political crises have stalled the plans for a long time. Given the present atmosphere of relative political and economic stability, Ethiopia hopes to expand her agricultural and irrigation capacities

(Sudan Update, 15 July 1995). The new international image has brought further confidence among the Ethiopian policy makers to develop their internationally shared water resources. Ethiopia is internally peaceful and her relationships with Western nations have improved considerably in recent years.

There is a growing confidence in Addis Ababa to successfully counter the Egyptian and the Sudanese objections to its water development projects at the diplomatic level. Egypt has already been able to exploit most of the irrigation and hydropower potential within its own borders. Countries like Ethiopia and Sudan are very much behind in this race. Economic and also political difficulties in Ethiopia and Sudan have so far hampered development of water resources, such as construction of dams and other water projects on the Blue Nile tributaries and in the Sudan. But these countries are gradually facing an increased demand for water and they want to develop and use the waters of the Nile tributaries. While Egypt opposes any agricultural water diversion in the upstream, it considers hydropower facilities in the Nile system are beneficial to its interest. The Sudanese dams built in the post-1959 Agreement period have acted as "siltation basins", stopping most of the sediment load before it reaches Lake Nasser. Though the blockage of sediments in Sudan has resulted in riverbed erosion in Egypt, it has protected the carrying capacity of Lake Nasser. Thus, Egypt encourages hydropower development in the upstream, particularly in Sudan, and it does not object to similar projects in other upstream countries.

As can be seen in Table 8.2, Egypt has developed almost all of its possible hydropower potential. Sudan has been able to exploit its hydropower potential to a small extent, while the other major riparian, Ethiopia, is far behind. The hydropower facilities do not decrease the water flow, while they help to reduce the sediments reaching Lake Nasser. Egypt also expects that hydropower can help Sudan and Ethiopia to utilize their groundwater potential to meet the increasing water needs, and that will ensure the long-term smooth passage of Nile water to Lake Nasser.

Table 8.2 Hydropower potential of Nile basin countries

Country	Hydropower potential (MW)	Installed (MW)
Burundi	1366	36
D.R. Congo	530,000	2829
Egypt	3210	2825
Eritrea	–	–
Ethiopia	162,000	378
Kenya	30,000	611
Rwanda	3000	59
Sudan	1900	225
Tanzania	20,000	339
Uganda	10,200	155

Source FAO (1997)

Escalating Conflict Scenario

In the 1990s, the Nile basin countries were engaged in open conflict over the water sharing issues. The establishment of the Nile Basin Initiative (**NBI**) in 1999 brought a certain shift to the bilateral relationships among the basin countries in general. Being part of the **NBI**, basin states expected substantial investment in the Nile basin. The Nile Basin Initiative is more than a decade old now. Many funding agencies, particularly the World Bank, projected the NBI as an unprecedented success story of river basin cooperation. It is true that this initiative brought almost all of the basin countries together and it eased riparian tensions. However, the NBI was expected to foster socioeconomic development in the basin by bringing the joint benefit sharing of the Nile water and taking adequate long-term ecological sustainability into account. On these fronts, very little has been done on the ground for which the **NBI** can take the credit. The state-centric perspective of Nile water development still dominates the basin.

To begin with, none of the Nile basin countries is yet to ratify the Convention on the Law of the Non-Navigational Uses of International Watercourses. The Convention stresses that states should use watercourses in an equitable and reasonable manner, and defines a procedure to follow when planned schemes may have adverse impacts on other riparian states. It is true that the ratification of this Convention will not be able to resolve the Nile water-sharing disputes altogether. However, the lack of support to it from the basin countries portrays the prevalence and domination of an old school of thought in the Nile water management sector. The NBI, due to its strategy of securing the consensus of all riparian countries, has made some progress on less controversial issues, but has postponed the important ones, particularly water-sharing matters involving Ethiopia, Sudan, and Egypt.

Moreover, the **NBI**'s almost total dependence on external funding for its inception and operation raises doubts about its long-term sustainability. The World Bank has worked as a facilitator to bring the basin countries together in multilateral talks leading to basin-based initiatives. Many developed countries have provided initial support and have promised to help with future initiatives. However, there is a clear absence of commitment of the basin countries to achieve satisfactory progress in the area of basin-based water management.

After years of meetings and deliberations, in June 2007, the Nile Council of Ministers drafted the Cooperative Framework Agreement (**CFA**), in which they expressed their desire for establishment of a permanent river basin commission. Though basin countries promise basin-wide cooperation, they continue to advocate and promote large-scale hydroprojects unilaterally within their own territories. Furthermore, the basin countries have not taken any measure to reduce their dependence on the Nile River water; rather, their demand for water is consistently increasing.

Recently, the basin-based Nile Basin Initiative has come to a breaking point. The **CFA** was expected to be signed in September 2007, and would have replaced the existing water agreement between Egypt and Sudan. However, Egypt and Sudan are opposed to the **CFA**, as they are apprehensive that it would take away their historical priority over the Nile water. While the **CFA** asks for the equitable and reasonable water use, **it** prevents harm to prior water users. It stands for the protection and conservation of ecosystems and the basin, while it makes prior informed consent mandatory for new water use. All these important articles are in line with the 1997 Convention on the Law of the Non-Navigational Uses of International Watercourses. The real disagreement surfaces on it, which is about water security.

Cooperative Framework Agreement (**CFA**) states: "Having due regard for the provisions of it, Nile Basin States recognize the vital importance of water security to each of them. The States also recognize that cooperative management and development of the waters of the Nile River system will facilitate achievement of water security and benefits. Nile Basin States therefore agree, in a spirit of cooperation, (a) to work together to ensure that all states achieve and sustain water security, and (b) not to significantly affect the water security of any other Nile Basin State." Though the upper riparian countries are in favor of this formulation, Egypt and Sudan demand section (b) of the article to be changed as "Not to adversely affect the water security and current uses and rights of any other Nile Basin States". The introduction of the "water security" concept in the CFA has brought a serious impasse and threat to its future, as the upper riparian countries vehemently reject the Egyptian-Sudanese proposal for amendment (Mekonnen 2010).

Several attempts to find a common ground have failed. In April 2010, Egypt and Sudan unsuccessfully introduced a new proposal seeking to establish the Nile Basin Commission before reaching final agreement on the **CFA**. In May 2010, seven upper riparian countries came together in favor of going ahead with the Cooperative Framework Agreement, ignoring objections from Egypt and Sudan. Ethiopia, Uganda, Tanzania, Rwanda, and Kenya have already signed this Agreement, which is rejected by the two powerful downstream, riparian countries, Egypt and Sudan. In June 2010, even Sudan demanded the freezing of all the Nile Basin Initiative Projects as a mark of protest against this upstream riparian initiative. The main reason for opposition by Egypt and Sudan is that the **CFA** does not protect their historic rights over the Nile water. The agreement may soon be signed and ratified by the other two upper riparian countries, Democratic Republic of Congo and Burundi, and this will pave the way for the establishment of the Nile River Commission. However, without Egypt and Sudan, the Commission will not be in any way capable of addressing the water sharing issues of the Nile River; rather, it may be a major source of contention between upstream and downstream riparian states.

The increasing threat of global climate change has brought further insecurity to the countries in the Nile basin about their future water demand and supply, and this has pushed them to initiate and expedite their own water projects. In spite of the international community's support for cooperative water management of the Nile water, it has not been able to take a foothold, and most of the basin countries, particularly Ethiopia, Sudan, and Egypt, have undertaken unilateral actions to protect their water interests. The changing geopolitical situation has provided alternative possibilities for Ethiopia and Sudan to raise financial and technical support for their own water development projects. Meanwhile, Egypt has lost some of its bargaining power, as it does not have the same level of influence internationally as it used to. Both these factors have dashed any hope of achieving basin-based management in the near future, and at the same time have increased the possibilities of open dispute among the riparian countries.

Fear of Climate Change

As climate change can potentially change water supply and demand patterns in the basin, sharing of the already scarce water resources of the Nile River may become a serious security challenge in the near future. Climate science has been able to provide an overall understanding of how the hydrological cycle will change at the global level, but the predictions of water demand and supply at regional and basin levels are not easy (Di Baldassarre et al. 2011). There is general consensus that the projected impacts of global climate change over freshwater may be huge and dramatic, but the impacts will not be on the same scale in each and every geographical region. Even within a river basin, the effects will vary depending on the location. This further increases the uncertainties and anxieties over the water availability in the Nile basin. Unfortunately, in the Nile basin, like many other international river basins in the Southern Hemisphere, "both domestic and international water laws and policies are inadequate to meet the challenges posed by this global phenomenon or to adapt to the additional consequences that appear to be inevitable" (Eckstein 2010).

The Intergovernmental Panel on Climate Change reports that up to 2×109 people may face water scarcity by 2050 (IPCC 2007). Most of the areas covered by the Nile River basin are projected to become warmer during this century, increasing the demand for freshwater. On the supply side, there are quite a few question marks over water availability. There is still lack of consensus about the projected changes in the basin's climatic means and extremes (Conway 2005). Due to substantial inter-model differences of precipitation, quantitative estimates of projected water supply changes are not easy to determine in an exact manner. There is also the possibility of local climate changes making it further difficult to assess a basin-wide trend. However, there is a strong likelihood that the climate change is going to multiply the uncertainty factor of the Nile River flow and may bring steady and significant reduction to it.

Global climate change may also increase the frequency of extreme events, such as droughts and floods in the Nile basin. The reduced water supply in the dry season is going to facilitate seawater intrusion to the river system, threatening agriculture, fisheries, and ecology in the delta (Wahab 2005). Temperature rises might reduce the productivity of major crops, and increase their water requirements. The Nile basin is also going to be severely affected by the impacts of climate change. The entire basin may receive increased precipitation early in the century, followed by decreases later in the century. This will also be reflected in the annual average Nile

River runoff, which is projected to increase from 11 to 14% in the first half of this century, and decrease 7–16% in the second half (Beyene et al. 2010). Thus, the agricultural sector in the basin, in particular, will experience increasing shortfalls by the mid- to late-twenty-first century. Furthermore, the impact of climate change can potentially bring considerable variation to the stream flow reaching Lake Nasser, creating further complications between Egypt and Sudan over water sharing.

The Nile water sharing agreement of 1959 between Egypt and Sudan is based upon the assumption that any resulting shortages will be for a short duration only. The treaty's "drought" provision is supposed to address short-term shortages with a temporary reallocation method (Tarlock 1999/2000). This has not been possible during the 1980s water shortages, as officials of both countries failed to agree on a new figure (Goldenman 1990). If Egypt and Sudan fail to agree on short-term cutbacks, it does not bode well when climate change can potentially bring long-term increases or decreases to the average run-off of the Nile River system. Moreover, climate change can further influence the sharp variability of the Nile water flows, which can possibly pose serious challenges for the water management in the basin. As global climate change might bring long-term changes to the volume and pattern of run-off in the Nile River systems, it is crucial to assess the quality and capability of on-going sharing arrangements to address this challenge. Climate-related changes require comprehensive adjustments in the on-going water management structure of the Nile River (Gleick 1988; Cooley et al. 2009). This comprehensive effort might ask for the water sharing arrangements to be flexible and adaptable in allocating reduced and surplus water flow, maintaining a certain water quality level, sustaining ecosystems, controlling flood, and protecting existing water development infrastructures. Thus, the river sharing arrangements need to have provision for information sharing, conflict management mechanisms, flexibility to adjust to uncertainties, and endeavor for basin-based development strategy (Goldenman 1990).

The Nile basin countries must be under obligation to regularly exchange data and information among themselves to be able to monitor and manage changing conditions affecting shared water. Even the 1959 Agreement between Egypt and Sudan does not have any obligatory provision to share hydrological data. In case of any dispute or disagreement over shared water management, there must be provision available to basin countries to manage them as soon as possible. Water management regimes and institutions in the Nile basin require a flexible mandate to plan, operate, and implement, in order to cope with changing climatic conditions. Mitigating or adaptive actions by an individual state to address potential climate change impacts in an international river basin are unlikely to achieve the objective. The emerging unprecedented situation due to changes in climatic patterns requires the Nile basin countries to cooperate and act collectively and jointly (Swain 2011).

In the Nile basin, the agreement among the disputing lower riparian countries to constitute the Nile Basin Initiative in 1999 was certainly a right step toward basin-based water management. However, more than a decade has passed, and no concrete progress has been made. In reality, very little progress has taken place to establish effective and cooperative water management institutions in the basin. Most of the riparian countries seriously continue to pursue large-scale unilateral dam construction. The international community, particularly the World Bank, has been claiming the credit since 1999 for creating a platform for a basin-based water cooperative framework; however, the on-going stand-off between Egypt and Sudan with the upper riparian countries over an article of the Cooperative Framework Agreement shows shallowness in the claim. In fact, the Nile basin is far from achieving a basin-based water management institutional structure (Swain 2011).

The so-called basin-based cooperative framework in the Nile was initiated with the active involvement of international donor agencies, and it still survives with their help and assistance. Even after more than a decade of existence, the **NBI** receives still very little contribution from the basins states themselves, exposing the lack of interest of the basin states toward joint management of the shared river resources. It is true that the Nile Basin Initiative was able to hide water-based incompatibilities in this highly volatile region for over a decade. However, the present form of water sharing and the quality of cooperation is not enough to meet the new challenges that might arise from climate change. These existing arrangements need to evolve and do so fast, constructively and positively. Unfortunately, there is very little sign of that in the Nile basin; rather, there have been movements in the opposite direction. The arrival of China as another presence in power politics has added a new impetus to the unilateral approach (Swain 2011).

The China Factor

The end of the last century was marked by some dramatic changes in power relations in North-East Africa. The Upper Nile riparian countries decided to come together and coordinate their strategy to challenge the historical domination of the Nile water, mainly by Egypt and, to some extent, by Sudan. In the earlier decades of the Cold War period, Egypt had used Soviet connections very effectively to obtain financial and technical support to capture a large share of the Nile water. After the Camp David Accords in 1978, Egypt received strong American backing to maintain its hydro-hegemony in the region. However, in this century, the power relation in Africa has transformed dramatically due to the arrival of China as a major player in the region. China and its economic might have provided the possibility of securing alternative external support to the Upper Nile riparian countries for large water development projects. Particularly, Sudan and Ethiopia have begun to exploit this opportunity to its full extent, and this has contributed in reducing their comparative economic and political weaknesses vis-à-vis Egypt. Egypt has been caught in a very difficult situation, as the support of the USA is not enough anymore to meet this challenge. However, it is using all its power and influence to delay the process of reaching an equitable way of water sharing of the Nile River.

China's investment in Africa was close to US$5 million annually in 1991, but by 2006 it jumped up to US$1.25 billion. Africa's trade with China has also been growing with an equally phenomenal speed. In the 1980s, trade between China and

Africa totaled only US$12 million per year; by 2000, it had reached US$10 billion, and by 2006 a staggering US$55 billion (Amosu 2007). Most of the African countries, particularly the Upper Nile riparian states, have abundant natural resources but lack the capacity to process and market those resources, providing a perfect opportunity for China. The Chinese involvement has become quite strong, particularly in Sudan and Ethiopia due to its primary interest in their petroleum, raw materials, and markets.

China's policy of non-interference in the domestic affairs of its African collaborators fits well with the African ruling elites. They prefer to deal with Chinese companies compared to the European nations and the USA, who insist on human rights and democratic reforms. China has provided African countries an opportunity to become free from the clutches of the West and Bretton Woods institutions. That has allowed China to expand its business in areas of Africa, such as Sudan and Ethiopia. The Rwandan President, Paul Kagame, who is also the chairman of the East African Community, in an interview to the German business daily Handel blatt, in October 2009, generously praised the Chinese investment in Africa: "Huge Chinese investment in African companies and infrastructure is helping Africa develop. The Chinese bring what Africa needs: investment and money for governments and companies. European and American involvement has not brought Africa forward" [**The National (Abu Dhabi), 9 January 2010**, http://www.thenational.ae/business/energy/chinasglobal-quest-for-oil].

China's growing economy requires massive levels of energy and minerals to sustain its spectacular growth. China is actively seeking to rapidly increase its oil import from Africa (Hanson 2008). Though the reported figure on Chinese aid and investment in Sudan was already more than US$4 billion between 2002 and 2007, the real figure may be even much more than that. According to a Sudanese diplomat, the total amount of the Chinese investments in Sudan in the past in the petroleum sector alone had reached more than US$6 billion (**Sudan Tribune, 6 November 2007**). The China National Petroleum Corporation (**CNPC**) started its investment in Sudan's oil sector in 1995 and, at present, it is the dominant actor in the country (Hellstrom 2009). The Chinese involvement in Sudan is not limited to the petroleum sector alone. Sudan receives Chinese support in three major project areas: oil refineries, highway construction and large dams (Mirak-Weissbach 2008). With Chinese financial and technical support (the **CCMD** consortium), Sudan has already constructed the Merowe Dam project on the Nile system. Sudan's President Omar al-Beshir inaugurated this large hydroelectric dam project in March 2009, the biggest water project on the Nile since the construction of the Aswan High Dam. This dam is primarily a hydropower project with potential capacity of 1250 MW, but Sudan expects to use the dam also for irrigation purposes in the future. Since 2008, Sudan has also engaged two Chinese state-owned firms, Sino hydro Corp and CWE, to raise the height of the Roseires Dam in eastern Sudan by an additional 10 m to increase the irrigation and power generation potential of the structure (**Sudan Tribune, 28 April 2008**). At present, these two projects are built within the Sudanese share of water according to the 1959 Agreement allocation, but the future dam building plans of Sudan raise serious worries for Egypt. Sudan also created a new official agency, the Dam Implementation Unit (**DIU**). The DIU has worked in strong cooperation with China since 2001, both on the Merowe Dam construction and on the heightening of the Roseires Dam, and also has plans to build the Kajbar and Dal dams on the Nile. Moreover, Sudan, through its state-owned agencies as well as private companies, is expanding its irrigated areas close to the banks of the Blue Nile and White Nile through pumping projects. Like Sudan, Ethiopia also sees China as an alternative to the West. Ethiopia, which was waiting long for foreign capital and technology to develop its water resources for hydropower production and irrigation, has accepted the Chinese support with open arms. The African Development Bank, the World Bank, and the European Investment Bank, due to open or clandestine lobbying by Egypt and Sudan, constantly refused to support several planned water projects in Ethiopia. However, China is now generously providing both financial and technological support to Ethiopia to undertake several of these projects, particularly large dams. China has become a major player in Ethiopia in infrastructure development and construction. With Chinese help, Ethiopia plans to increase its hydropower generation, to a large extent for its own use, as well to export energy to neighboring Kenya, Sudan, and Djibouti.

With Chinese support, Ethiopia has already built the Tana Beles modern hydropower dam complex. This 460-MW dam on the Nile was inaugurated on May 14, 2010, the same day as the new Cooperative Framework agreement was signed in Entebbe. The Tana Beles project is primarily focused on hydropower production; however, the Ethiopian Ministry of Water Resources claims that it is designed to be an integrated Water Resources Development Project to provide irrigation to a total area of more than 100,000 ha of land (Ministry of Water Resources 2010). The Ethiopian Government has already engaged several Chinese firms to develop the rich hydropower potential of the country. In November 2009, Ethiopia, in collaboration with China, completed the construction of a 188-m-tall hydropower dam on the Tekeze River, a tributary of the Nile. The China Water Resources and Hydropower Engineering Construction Company and the National China Gezhouba Construction Group Corporation for Water Resources and Hydropower were involved in the construction of this project (**Addis Fortune**, **13 June 2010**). Ethiopia has an estimated hydropower potential of 45,000 MW, ample for her own demand with much to spare for exports. With Chinese support, Ethiopia not only aims to increase its hydropower production but also has successfully started contesting the existing status quo over the use of the Nile water. Besides hydropower, Ethiopia also seeks to improve her agricultural production with Chinese support. As the Ethiopian Foreign Minister explained in January 2008, “China is an important partner for us. It is involving in the area of our priorities such as agriculture, infrastructure including roads rural hospitals and schools” (**The Daily Monitor**, **12–13 January 2008**). Moreover, Chinese companies have also recently acquired large agricultural land in the Ethiopian part of the Nile basin for farming purposes. Besides Ethiopia and Sudan, China is also engaged in providing support for dam building in other Nile riparian countries, such as Uganda, Burundi, and the Democratic Republic of Congo (Brewer 2008).

The Riparian Countries of the Nile Understanding

The Nile River offers a limited amount of water for nearly 400×10^6 people living within its basin countries today. Increasing population and economic pressures in these countries are mounting faster than the river's capacity to support. With the growing multiple water demands, in combination with the high spatial and temporal variability of water availability, the necessity for cooperation and coordination among the riparian countries in the Nile basin becomes a crucial issue. Political leaders, policy analysts, and researchers consistently argue that cooperation on development and management of the Nile water resources can yield major benefits from the river on food and energy production, and will foster many other benefits for the development and well-being of the basin population.

At the end of the 1990s, the riparian countries of the Nile, prompted by the Western donor community, came into an understanding to find ways and means to achieve basin-based cooperation. Though more than a decade has passed, that dream has not yet been realized. In the face of mounting pressure from the upstream countries, Egypt and Sudan are working hard to maintain their historic rights over the Nile water. Ethiopia is trying whatever possible to capture its share of the Nile water, which has been denied to it in the past due to its own political and economic weaknesses. As Ethiopian Prime Minister, Meles Zenawi, argues that "The current regime cannot be sustained. It's being sustained because of the diplomatic clout of Egypt. There will come a time when the people of east Africa and Ethiopia will become too desperate to care about these diplomatic niceties. Then, they are going to act" (**The Columbus Dispatch 29 May 2010**). Most of these upstream Nile riparian countries have already acted by signing the Cooperative Framework Agreement in May 2010, in spite of strong opposition from Egypt and Sudan.

A large part of the Nile basin is considered as one of the poorest regions of the world. Water scarcity is a major challenge for this already closed basin. The challenge is further exacerbated by climate variability. Thus, the immediate national interests of the riparian countries are taking priority over the basin-based strategy. The Western donor community, which was behind the initiation of basin-wide cooperation, is rapidly losing its power due to the growing presence of a new economic giant in the region, China. The possible dissolution of Sudan in 2011, after a Southern Sudanese independence referendum, might bring in further complications to the Nile water sharing. Oil fields in south Sudan will attract Chinese investment and that will make the north more vulnerable, economically as well hydrological. In anticipation of this eventuality, the regime in Khartoum has started collaborating closely with its Arab neighbor, Egypt, and is resolutely against the Common Framework Agreement in its present form. After a decade of failed attempts to initiate cooperation, the countries of the Nile basin have again started adopting conflicting postures over the water. It is necessary for the important riparian states of the Nile basin to abandon their state-centric water development approach and develop sustainable cooperation over the shared water to meet the climate change challenges (Swain 2011).

References

Agreement. (1959). *Agreement between the Republic of the Sudan and the United Arab Republic for the Full Utilization of the Nile Waters* [1959 Agreement], Cairo (8 November 1959). Available at http://www.fao.org/docrep/w7414b/w7414b13.htm. Last visited 27 April 2011.

Amosu, A. (2007). *China in Africa: It is (still) the governance, stupid*. Washington, DC: Institute for Policy Studies.

Beyene, T., Lettenmaier, D. P., & Kabat, P. (2010). Hydrologic impacts of climate change on the Nile River Basin: Implications of the 2007 IPCC scenarios. *Climate Change, 100*(3–4), 433–461.

Blackmore, D., & Whittington, D. (2008). An Independent Report of the Scoping Study Team to the Eastern Nile Council of Ministers, final report, Opportunities for cooperative water resources development on the Eastern Nile: Risks and rewards, October 2008.

Brewer, N. (2008). *The New Great Wall—A guide to China's overseas dam industry*. Berkeley, CA: International Rivers Network.

Brunnee, J., & Toope, S. J. (2002). The changing Nile basin regime: Does law matter? *Harvard International Law Journal, 43*, 105, 122–123.

Christiansen, S. (1992). Shared benefits, shared problems. In S. Lodgaard & A. H. Ornäs (Eds.), *The environment and international security* (p. 57). Oslo: PRIO Report no. 3.

Clarke, R. (1991). *Water: The international crisis*. London: Earthscan.

Collins, R. O. (1990). *The waters of the Nile: Hydro politics of the Jonglei Canal, 1900–1988*. Oxford: Clarendon Press.

Conway, D. (2005). From headwater tributaries to international river: Observing and adapting to climate variability and change in the Nile basin. *Global Environmental Change, 15*, 99–114.

Cooley, H., et al. (2009). *Understanding and reducing the risks of climate change for trans boundary waters*. Oakland: Pacific Institute.

Dawoud, K. (2001). Taming the Nile's serpents: Conflict over water supply. Paris: United Nations Educational, Scientific and Cultural Organization (UNESCO).

Di Baldassarre, G., et al. (2011). Future hydrology and climate in the River Nile basin: A review. *Hydrological Sciences Journal, 56*(2), 199–211.

Eckstein, G. (2010). Water scarcity, conflict, and security in a climate change world: Challenges and opportunities for international law and policy. *Wisconsin International Law Journal, 27*(3), 409–461.

El-Moattassem, M., Mercer, A. G., & Mayor-Mora, R. (1996). A strategy for protection, development and preservation of the Nile River. In M. E.-M. A. M. Shady, E. A. Abdel-Hafiz, & A. K. Biswas (Eds.), *Management and development of major rivers* (pp. 28–42). Calcutta: Oxford University Press.

El Quosy, D. E. A. (2001). *Agricultural development in Egypt across two millenniums*. Paper presented to the World Development Federation Virtual Global Super Projects Conference (November).

Fabunmi, L. A. (1960). *The Sudan in Anglo—Egyptian Relations: A Case Study in Power Politics, 1800–1956* (p. 147).

FAO (Food and Agriculture Organization). (1997). *Irrigation potential in Africa, a basin approach*. Rome: FAO of the United Nations.

Farrag, A. H. A. (2005). The hydraulic and hydro chemical impacts of the Nile system on the groundwater in upper Egypt. *Assuit University Bulletin Environment Resources, 8*(1).

Gardner-Outlaw, T., & Engelman, R. (1997). Sustaining water, erasing scarcity: A second update. Washington, DC: Population and Environment Program, Population Action International.

Gleick, P. H. (1988). The effects of future climate changes on international water resources: The Colorado River, the United States, and Mexico. *Policy Science, 21*, 23–39.

Gleick, P. H. (1993). Water and conflict. *International Security, 18*(1), 79–112.

Goldenman, G. (1990). Adapting to climate change: A study of international rivers and their legal arrangements. *Ecology Law Quarterly, 17,* 741–802.
Hanson, S. (2008). *Backgrounder: China, Africa, and oil.* New York: Council on Foreign Relations.
Hassan, H. A., & Rasheedy, A. A. (2007). The Nile River and Egyptian foreign policy interests. *Africans Social Revolution, 11,* 29.
Hellstrom, J. (2009). *China's emerging role in Africa.* Stockholm: FOI.
Homer-Dixon, T. F. (1994). Environmental scarcities and violent conflict: Evidence from cases. *International Security, 19*(1), 5–40.
Hvidt, M. (1995). Water resource planning in Egypt. In E. Watkins (Ed.), *The Middle Eastern Environment* (pp. 90–100). Cambridge: St Malo Press.
IPCC (Intergovernmental Panel on Climate Change). (2007). Climate change 2007: The physical science basis. In S. Solomon et al. (Eds.), *Contribution of working group I to the fourth assessment report of the intergovernmental panel on climate change.* Cambridge, UK: Cambridge University Press.
IUCN, IWMI, Ramsar, & WRI. (2003). Water resources atlas. Length of the Amazon River is adjusted according to the new claims made in 2007.
Kendie, D. (1999). Egypt and the hydro-politics of the Blue Nile. *Northeast African Studies, 6,* 141, 145.
Kliot. (1958). The 1929 Agreement as a culmination of Anglo-Egyptian hegemony over the Sudan which Sudan, thus, did not consider itself bound by and eventually abrogated it in 1958.
Koutsoyiannis, D., Yao, H., & Georgakakos, A. (2008). Medium range flow prediction for the Nile: A comparison of stochastic and deterministic methods. *Hydrological Sciences Journal, 53*(1), 142–164.
Mason, S. A. (2004). *From conflict to Cooperation in the Nile Basin* (Ph.D. Thesis). Swiss Federal Institute of Technology, Berne, Switzerland
McCaffrey, S. C. (1993). Water, politics, and international Law. In P. H. Gleick (Ed.), *Water in crisis: A guide to the world's fresh water resources* (p. 94). New York: Oxford University Press.
Mekonnen, D. Z. (2010). The Nile Basin Cooperative Framework Agreement negotiations and the adoption of a Water Security Paradigm: Flight into obscurity or a logical cul-de-sac? *European Journal of International Law, 21*(2), 421–440.
Mekonnen, D. Z. (2011). Between the Scylla of Water Security and Charybdis of Benefit Sharing: The Nile Basin Cooperative Framework Agreement—Failed or just teetering on the brink? *Gottingen Journal of International Law, 3*(1), 345–372.
Ministry of Water Resources. (2010). Federal Democratic Republic of Ethiopia. http://www.mowr.gov.et/index.php?pagenum=0.1&ContentID=42. Viewed 1 October 2010.
Mirak-Weissbach, M. (2008). *A new Asian-African alliance? China support's Sudan's economic growth.* Montreal: Center for Research on Globalization.
Mohammed N. A. L. (1996). *Environmental conflicts in Africa.* Paper presented to NATO advanced research workshop on conflict and the environment 12–16 June, Bolkesjø, Norway.
Nassar, G. (1995). War of Words and Water, Al-Ahram Weekly, 6 July.
Okidi, C. O. (1980). Legal and Policy Regime of Lake Victoria and Nile Basins. *Indian Journal of International Law, 20,* 395, 423.
Okidi, C. O. (1996). *Legal and policy considerations for regional cooperation on Lake Victoria and Nile River.* Paper presented to The Nile 2002 Conference. February, Kampala, Uganda.
Postel, S. (1995). Where have all the rivers gone? *World Watch, 8*(3), 12.
Rangeley, R., Thiam, B. M., Andersen, R. A., & Lyle, C. A. (1994). International River Basin Organizations in Sub-Saharan Africa, World Bank Technical Paper no. 250. New York: The World Bank.
Sehmi, N. (1996). The enigmatic Nile. Geneva: World Meteorological Organization. *WMO Bulletin, 45*(3), 267.
Smith, S. E. (1986). General impact of Aswan High Dam. *The Journal of Water Resources Planning and Management, 112*(4), 551–562.

Sofer, A. (1999). *Rivers of fire: The conflict over water in the Middle East.* Lanham: Bowman & Littlefield.

Sudan Update. (1995, July 15). Water as a Weapon. *Sudan Update, 6*(1). http://www.africa.upenn.edu/Newsletters/SDate611.html

Suliman, M. (1992) Civil war in Sudan: The impact of ecological degradation. Zürich & Bern: Environment and Conflict Project, Occasional Paper no. 4, 21.

Swain, A. (1996). Water scarcity: A threat to global security. *Environment & Security, 1*(1), 156–172.

Swain, A. (2002). The Nile Basin Initiative: Too many cooks, too little broth. *SAIS Review, 22*(2), 293–308.

Swain, A. (2008). Mission not yet accomplished: Managing water resources in the Nile River Basin. *Journal of International Affairs, 61*(2), 201–214.

Swain, A. (2011). Challenges for water sharing in the Nile basin: Changing geo-politics and changing climate. *Hydrological Sciences Journal, 56*(4), 687–702.

Tarlock, A. D. (1999/2000). How well can international water allocation regimes adapt to global climate change? *Journal of Land Use & Environment Law, 15*, 423–449.

The Republic of Sudan. (1996). *Country Paper, Paper presented to The Nile 2002 Conference.* Uganda, February: Kampala.

UNEP, DEWA, & GRID. (2000). *Water Sharing in the Nile River Valley, PROJECT GNV011.* Nairobi: UNEP.

Wahab, H. M. (2005). *The impact of geographical information system on environmental development* (M.Sc. Thesis). Al-Azhar University.

Warburg, G. R. (1991). The Nile in Egyptian-Sudanese relations. *Orient, 32*(4), 570.

Warbung, G. (2000). The Nile in Egyptian-Sudanese Relations, 1956–1995. In H. Erlich & I. Gershoni (Eds.), *The Nile: Histories, cultures and Myths* (pp. 229–230).

Wolfensohn, J. (2001). Rich nations can remove world poverty as a source of conflict. International Herald Tribune, 6 October.

Chapter 9
Agriculture and Irrigation in Egypt

Abstract This chapter provides a brief overview of the history of agriculture and irrigation in Egypt, background of agriculture and irrigation in Egypt, scope of the chapter, cooperative relations, conflicts and divergent perceptions, Egypt and Ethiopia behind the renaissance dam, doctrines, principles and practices of international law, doctrines and international conventions, principles and rules.

Keywords Scope of the chapter · Cooperative relations · Conflicts and divergent perceptions · Egypt and Ethiopia behind the renaissance dam · Doctrines · Principles and practices of international law · Doctrines and international conventions · Principles and rules.

International Law, in general, exists and effectively contributes to positively settle a wide range of controversial issues. Despite that, it also equally has several more loopholes to strictly examine and mend particularly in major areas of International Water Law. Due to its inherent nature as a rule applying to the most complex interstate relations, international water law obviously could not be as much compressive, to the point and accurately instrumental as required and desired. With particular reference to international rivers, international law undeniably is too general while international rivers are too particularistic. International law tends and seriously tries to be too objective regardless of the unavoidable particularity of each state in question, state interest in international rivers is too subjective. International law tends to be too vague, ambiguous, and idealistic in its provisions and statements, while state use of international rivers demands an approach which is as clearer as possible, straightforward, and realistic. The Ethiopian Renaissance Dam represents the fact that state practice in the specific area of international waters is still guided by state practice usually through unilateral actions. As such, international principles, in this case, follow far behind state practice, in the political sense of the term (Alebachew 2011).

Ethiopians now stand as a legal challenger not only to the timely relevance of the traditional Egyptian policy that founded itself on the perceptions of Ethiopia's capacity limitations to make use of the Nile waters but also to the adequacy

© Springer International Publishing AG 2017
Y.M. Hamada, *The Grand Ethiopian Renaissance Dam, its Impact on Egyptian Agriculture and the Potential for Alleviating Water Scarcity*,
Environment & Policy 55, DOI 10.1007/978-3-319-54439-7_9

of international law to preempt interstate misunderstandings. Obviously, the Renaissance Dam has showed that Egyptians have created a formidable duty more on themselves than on Ethiopia by their insistence on pursuing "No Negotiation" Nile Policy. At present, it means that Egyptians, in demonstrating their loyalty to their policy, have to wait patiently until practice proves whether the Dam would actually harm or does not harm their advantages. Legally speaking, Egypt finally finds itself prisoner of its own policy (Alebachew 2011).

Background

"The Renaissance Dam" (The Ethiopian Renaissance Dam 2011) is a new hydrological event in Africa, in general, and in Ethiopia, in particular. It is at present under construction at the very course of the Blue Nile that contributes about 86% of the total Nile waters. The specific site of the project is some 42 km away from the eastern border of the Sudan. The Blue Nile Dam under construction is reportedly one of the largest Dams in Africa, two times larger than the inland Lake Tana in Northern Ethiopia. The Dam will estimably hold about 63 billion L^3 of water at completion. The Ethiopian government designed to generate about 6000 megawatts of hydroelectric power. The hydropower planned to be generated from the dam is expected to raise the national power supply three times more than Ethiopia has now had so far. The project is estimated to cost about $5 billion or about 80 billion Ethiopian birr up to its completion after five or less years (Zenawi 2011a, b). A renewed academic discussion is currently high on the air in many places but more articulate and serious in Ethiopia and Egypt, to a lesser degree, in Sudan. This discourse is a significant point more for scholars of International Law for three main reasons. (A) The construction of the Dam was made official by the government of Ethiopia on April 4, 2011, while there were no any official agreements between Ethiopia versus Egypt and Sudan. (B) It was also launched at the time when Egypt and Sudan officially expressed that they had not so far found any reasonable ground to revise the 1959 Bilateral Agreement; (Egypt and Sudan agreement 1959); (C) It is fairly, however, simple to assume that the huge size of the Blue Nile Dam in Ethiopia will have a remarkable effect on the established perceptions in Egypt of water use and distribution that, many advices, needs some kind of Interstate Agreement.

This chapter is a systematic attempt at examining the backgrounds of interstate relations in light of Nile water utilization patterns up until the official commencements of the new Dam. It will also consider the backgrounds of the doctrines, principles, conventions, and customary laws which Ethiopia and Egypt have long been advancing in their respective rights, adversely affecting venues toward making working bilateral water agreement and the spill-over effect of the Dam on the international body of laws and rules on transboundary rivers and their utilizations.

Scope of the Chapter

The Ethiopian Dam is conceived and its official start has been announced amidst traditional interstate conflicts, first and foremost, between the major contributor of the Nile waters, Ethiopia, and Egypt, with latter being the major recipient country, followed by Sudan (Evans 1990). Before discussions on conflicting and cooperative relations among the three major Nile Basin states under view, it is therefore sound to disclose existing variations or patterns in the utilization of Nile Waters among the states. The Nile water in Egypt is widely used first for irrigated agriculture since from ancient times. With dramatic changes in global production structures from agriculture to manufacturing and industry especially since the turn of the eighteenth century in Europe and America. Egypt was able to become a major producer and supplier of cotton to American Markets (Knott and Hewett 1990). Following British colonization of Egypt, the introduction of improved production technologies, more water from the Nile River began to be used. The growth and expansion of urbanization in recent Egypt has generally followed the map of Nile where more water became the basis of urban life. Nile River in Egypt also furnishes inland transport rout, fishing, tourism, sports, and environment. Egypt completed the construction of Aswan Dam in 1902, enlarged it in 1912, and heightened in again 1934 to store 5.1 billion L of water. There were more than ten barrages extending from Aswan Dam making irrigated agriculture possible like Assuite, Zifta, Esana, Nag Hamuidi, Edifina (Abate 1990). The completion of High Aswan Dam, the Nasser Lake, in 1963 with a total storage capacity of about 162 billion L^3, and 107 billion L^3 of actual storage, made a whole year agricultural production possible. In a nutshell, Egypt's share in the use of the Nile waters has remarkably progressed to surpass its 55 billion L^3 legal quota anchored in the 1959 bilateral Agreement with Sudan. Sudan began utilizing Nile River for agricultural Irrigation purposes as early as 1910 while it was under British colonial rule. To date, Sudan has been producing a large amount of cotton and other cash crops (Tvedt 1992). Ethiopia's share until the Great Ethiopian Nile Dam was limited to only 0.6% of the total water. The Grand Abay Dam project of Ethiopia will be the second biggest utilization of Nile Waters next to Egypt and ahead of Sudan by catchments calculation. It is the first biggest hydroelectric power project to operate at a single location with the largest power potential but while the statistics is more than 110 in Egypt and 18 million ha in Sudan. 2 million tones silt from Ethiopia that sediment into the Nile water in Egypt causing a loss of about 2.9 billion L^3 of water will be reduced substantially by some 90% (Kendie 1999).

Cooperative Relations

Paradoxically enough, the “Ethiopian Renaissance Dam” has been made official amidst cooperative relations among Ethiopia, Sudan, and Egypt in other spheres of interstate relations (Zenawi 2011a, b). Ethiopia and Egypt have signed and implemented a dozen of bilateral agreements over joint and individual investment ventures in such areas as agriculture, manufacturing, trade and industry amounting to about $2.1 billion at the announcement of the construction of the Ethiopian Dam. One Egyptian mining company, for instance, congratulated the two governments that it has discovered a reliable store of gold just a few kilometers away from the very location of the Renaissance Dam just only after 5 days the Prime Minister of Ethiopia laid the Dam’s cornerstones (Weekly 2011a).

There has been a steady growth in the overall Egyptian direct foreign investment flow into Ethiopia with successful establishments of such factories like pipelines, textile, shoe and leather, to mention but only a few. There has been a joint interstate cooperation commission regularly meeting at the level of ministers between the two states and signed a lot of agreements in such areas as sports, education, culture, history, transport, information, technology, and science. In the sphere of international security and diplomacy, Egypt and Ethiopia have similar foreign policies. Egypt voted for the Head Quarters of the African Union 2004, at the Union’s General Assembly Meeting in Tunis, to continue to be seated in Addis Ababa at the time when Libya pressured African leaders to decide for their relocation in Tripoli (Zenawi 2007). Ethiopia and Sudan also have signed and implemented a series of interstate agreements on a wide range of issues like border security, control over terrorism, drug trafficking, etc. A number of Sudanese investors have established several manufacturing plants in many parts of Ethiopia like meat processing plants, mineral waters, textile, and others (Weekly 2011b). Sudanese investors and business people have been granted visa-free entry rights into Ethiopia’s territory whenever they need to do so. Ethiopia has been playing a role of a trusted referee, a mutually agreed Good Office. Sudan agreed for the deployment of entire military regiments from Ethiopia only, where Ethiopia contributed about 5000 strong to be deployed in Darfur but under United Nations’ Command and another 4300 along the border between southern and northern Sudan (Yenus 2011).

Conflicts and Divergent Perceptions

There have been similar patterns of conflicting interests despite all these cordial relations among Egypt, Sudan, and Ethiopia. Historically, it is recorded that Egypt under Ismail Khedive Pasha invaded northern Ethiopia and tried to control the Blue Nile from its source as a colonial territory in 1874 and 1876, at Gunda Gundi and Gurea receptively. These invasions resulted in total military reversals of Egyptian forces by Emperor Yohhanes IV of Ethiopia. Since then, there was no any move by

Egyptians to directly attack and control Ethiopia militarily in any degree (Ayele 2005). The major form and manifestation of conflict takes serious diplomatic and, at times, proxy interventions in the internal affairs of the other particularly between Ethiopia and Egypt. This conflict obviously emanates between them mostly as the result of their divergent hydro-topographic positions as suppliers and recipients respectively. At grassroots level, the relations particularly between the peoples of Egypt and Ethiopia have been grossly characterized by a shadow of mutual suspicions, hostile perceptions, conspiracy theories, and deeply entrenched emotions directly related to their positions to the Nile River.

At state level, Egyptian and Sudanese official foreign policy statements on Nile River are known more in practical manifestations than documented principles and strategies. Egypt particularly has long appeared to see the issue around Nile as something "permanently sealed" by working international laws and conventions (Arsano 2011a, b). The Foreign Relations and National Security Policies and Strategies of the Ethiopian Government, however, explicitly underscore that Egypt has long been trying to weaken Ethiopia in order to insure that its monopoly control and use of the Nile Waters would continue unrivaled and uncompeted (Alebachew 2011). Ethiopia complained over Egyptian "interventionist behavior" of assisting the separatist war in Eritrea before 1991 and the bloody border war of 1998–2000 after the latter's independence (Reta 2000). The most visible manifestation of Egyptian suspicion of Ethiopia occurred following the event in Addis Ababa in May 1997 when a group of terrorists attempted to kill the former president of Egypt, Hosni Mubarak. The event provoked a heated diplomatic squabble among Ethiopia, Sudan, and Egypt where, unexpectedly, the latter two shared the same position against Ethiopia. Ethiopia accused Sudan for sponsoring the terrorist attack and convinced the United Nations, which in 2000 voted for a 2-year military sanction against the Khartoum Government, which, Egypt, however, was opposed to by voting against the proposal. All in all, despite recent cooperative trends, the entire relations among these countries have been dominated by a fluctuating rise and fall of conflicts specifically with matters related to the Nile River.

Egypt and Ethiopia Behind the Renaissance Dam

The Ethiopian Renaissance Dam was made official almost unexpectedly for all pertinent parties to result in producing far reaching consequences on the doctrines, principles, and relevance of International Water Law as well as on the long-held policy traditions of the said states. One could study the dam's effect from various angles; however, it is enough here to identify the major background trends and events behind the dam with continued implications at present.

First, Ethiopia has, apparently by deliberate design, unusually been engaged in spearheading the coming into existence of the Nile Basin Initiative 2007, which consists of the seven Upper riparian states in addition to Ethiopia, namely Uganda, Kenya, Tanzania, Rwanda, Burundi, and Democratic Congo. The Initiative is supposed to be the first legal framework. While Sudan refrains from joining the Agreement, Egypt's response to the demand for signature has been an automatic rejection by, of course, the ousted regime of the ex-Egyptian president, Hosni Mubarak. Egypt recognizes no other international set of international water laws or agreement except the Egyptian-Sudan 1959 Full Utilization of Nile Waters Treaty (Guariso et al. 1987). Egypt consistently declined offers for active participation arguing that these are adverse to its "unquestioned right" to its share of the Nile waters as per the Treaty mentioned above.

Second, Egyptian leaders adopt a different form of regional cooperation in which they are willing to support Egypt-sponsored and monitored utilization of Nile Water projects in lower riparian states. This Egyptian policy orientations have grown up from surviving spirit borne out of a series of colonial Treaties: the Anglo-Italian Protocol of 1891, for the continued flow of Tekeze Waters the British Colony of Egypt through Sudan, the 1906 Tripartite Agreement among France, Italy, and Britain recognizing British political influence over the Ethiopian Territory of the Nile Basin, the 1929 Britain–Egypt Agreement recognizing Egypt's natural and historical rights over the Nile Waters, and many others (Gibb and Partners 1978). Egyptians were able to convince Uganda and Sudan to accept these terms in the 1980s. As the result, Uganda agreed for canceling a construction of additional water reservoir facilities at the mouth of Lake Victoria in the 1990s; (Addis Zemen Daily 2011).

Third, Ethiopian Government officials disclosed that they made the project official after they first exhausted all possible global sources of finance which were convinced by Egypt and Sudan not to assist the former. This led the Ethiopian people to embark on launching the project by insuring domestic sources of funding to wholly cover all costs. The apparent reason for Egyptian-Sudanese effective lobby was expressed to be the usual argument that any such project would harm their national interests. According to Egyptian sources, Egyptian authorities were not formally informed by the Ethiopian government about the dam. The Ethiopian government posed the argument that this would happen only through the Nile Imitative Framework (Kendie 1999). Moreover, Ethiopia announced the new Nile Project after successfully completing other similar hydroelectric projects at other Nile Tributary Rivers like Tana Belles and Tekeze, which did not entail heated debates.

Fourth, the Dam came into official announcement at the very time when Egypt was politically troubled by a revolutionary uprising. Sudan, too, was faced with rising instability engendered by the newly emerging Southern Sudan. Moreover, the Sudanese president was under persecution by the European Human Rights Court. Amidst all these, there came no new developments in the sphere of international water law that could affect the construction of the dam negatively or positively. Even so, the newly instated provisional government of Egypt first tended to insist on the traditional Egyptian Nile Policy that Ethiopia should discharge its duty of notifying Egypt of the project before launching construction. However, it remains questionable whether the readiness of Ethiopia to meet these preconditions would convince Egyptians to approach realistically the dispute. Egypt recently agreed only to name a Joint Technical Commission that would study the impacts of the project (Alebachew 2011).

These major backgrounds of the dam have induced debates. Many scholars of international law tend to emphasis a shift in a political balance of power among the three states behind Ethiopia's move to launch the dam's construction (Arsano 2011a, b). Other scholars dismiss this argument advancing the idea that what Ethiopia took as an opportunity was rather the weaknesses of international water law (Reda 2011). What does this mean? Ensuing sections attempt to clarify this.

Doctrines, Principles, and Practices of International Law

From politico-legal points of views, while the three states easily come to mutual understandings over being subject to general international principles of civilized states, they soon inter into fierce politico-legal squabbles over internationally renowned Doctrines, Conventions and Rules governing interstate relations. Let us see the matter orderly by starting discussion first on major doctrines developed to govern theories, principles, conventions, and rules of utilizing transboundary Rivers.

Doctrines

The first doctrine ever applied was "Absolute Territorial Sovereignty Theory" (Mc-Caffrey 1996) which argues that a state has complete control over all waters lying within its territorial jurisdiction, and may utilize those waters without regard to its effect on other co-riparian states, especially downstream states. This theory does not sound consistent with the demands of states under present legal views and moral thoughts, which uphold "mutual benefit" as a guiding principle. As such, both Ethiopia and Egypt have the least taste and inclination to this theory.

The second doctrine "Absolute Territorial Integrity" theory constitutes the opposite extreme of the Harmon Doctrine. According to scholarly arguments (Oppenheim 1955), a state is entitled to expect that the same volume of water, uninterrupted in quantity and unimpaired in quality, flows into its territory. Therefore, it argues that no interruption, diminution, or augmentation of the flow caused by the use of waters located in a riparian state is to be permitted. The upper-stream state under view has the right to exploit the waters of a river so long as such utilization does not affect the interests of the lower riparian. On the other hand, downstream states have the right to object any upstream water utilization that is likely to disrupt the natural flow or adversely affect the water position of the former. This again has the least policy taste particularly for Ethiopia as one and the major contributor of the Nile waters from among the upper riparian states. Egypt is also cold to the theory for largely pragmatic reasons that this would scarcely fit existing international realities.

The doctrine of "Community of Interest" (The newly launched Nile Basin 2009) argues that a state's sovereignty is limited by similar rights of other states sharing the same basin. This theory emphasizes the maximum utilization and optimum economic development of an entire river basin. It reinstates the community of interest approach and attributes a positive duty to render active cooperation in the rational development and utilization of the shared water resources. In other words, this theory considers the water of an international drainage basin to be managed as a unit without regard to national territorial boundaries. Therefore, the various co-riparian states should manage and develop the drainage basin jointly and share the benefits derived there from. This theory has stood as Ethiopia's kernel of foreign policy over the utilization of Nile River in the previous 15 years (Foreign Policy and National Security Policies' Document 2011).

This theory, on the Egyptian side, has never been widely accepted and applied, especially with regard to the non-navigational uses of international watercourses for three major reasons of top national interest: one is that the theory does not consider the different degrees of dependence among states on international waters; two, it does not consider the existence of alternative sources of water for upper riparian states; three and last, it does not consider differing capacities among states in utilizing international rivers, the MoF notes (Dingamo 2010). This theory was, however, upheld by the Permanent Court of International Justice in the adjudication of several cases at different times. The theory attracts Ethiopia more than Egypt as a basis of general principle and policy toward the utilization of the waters of the Nile River.

Limited "Territorial Sovereignty and Integrity Theory" (Caponera 1992), is the fourth doctrine. This approach holds that states have a sovereign right to use a water resource in international watercourses lying within their own territory to the extent that such use does not cause any injury or harm to other states. Accordingly, if the use of water resources in international watercourses by a state affects the interest of the co-riparian state, then the sovereign right of the state to the use of water in the international watercourses is limited. This theory is the entire policy foundation of Egypt (Bartnisky and Neicheko 1981) for a long period of time with a set of other conventions and state practices to satisfy its national interests. Despite this, some of the most important Egyptian international conventional and legal preferences are also shared partially by Ethiopia, as MoE notes. The two states diverge, however, on other key principles and details which make the difference far apart and visible. Now, let us see how these theoretical differences affect subsequent policy cleavages between Egypt and Ethiopia with elevated debates occurring due to the newly launched Nile Dam Project.

International Conventions, Principles, and Rules

The most frequently cited rules, conventions, and principles include the Helsinki Rules (1966); Rules of International Water Courses (1997), and the United Nations Convention (1997) on Non-navigational Uses of International Watercourses. Egypt banks on these international conventions for all its policy rationales toward Ethiopia's involvement in the utilization of the Nile River (Amr 2002). It argues that states should and could not "divert" the natural directions and courses of river flows, which would have always negative consequences on lower riparian states. "No-diversion" argument is the traditional top policy priority of Egypt. Egyptians broaden the scope of "diversion" to include many other development activities. The term "diversion" as defined may differ linguistically from other terms related to some acts which affect the flow of watercourses, such as extracting water and holding the flow of water. "Withdrawal" or "extracting" of water means that a State withdraws or extracts water from a watercourse flowing into another state, usually a neighboring country, and without the consent of the latter state (Helsinki Rules 1966). The best example of such an act is Israel's extraction of water from the Litany River in South Lebanon. The other term "holding or stopping the flow of water" means that any act taken by a riparian State in order to fill reservoirs behind dams. The best example for such an act is the Turkish act which stopped the flow of the Euphrates River to fill the reservoir behind the Ataturk Dam (Berber 1959).

For Egypt "diversion" over the River Nile or any of its tributaries is said to occur if the activities of the upper riparian states result in any one of the following changes on its water position: first, if it causes shortage of water supply as different from the amount previously used and held; second, if it causes damage to the environment; third, if it results in Reduction in the Level and Quality of the Groundwater; fourth, if it causes Shortage in Hydroelectric Power; and fifth, if it inflicts other impacts including like the interception of a large proportion of the sediment carried by that river and if the diversion might damage tourist sites of the lower riparian states (Arsano 1990). There are usually four traditional ways of interstate relations to implement such principles of international law and let us see which one Egypt opts for

A. states may sign a Treaty providing for a total prohibition of "diversion";
B. states may sign for a Treaty allowing "diversion" under total freedom of action;
C. states may sign Treaties over "diversion" of specific rivers, volumes, time schedules, tolerable degree of damages, etc.; and,
D. states may sign no Treaty at all leaving "diversion" decisions to be based on accepted "principles of international law" (Arsano 1999). Scenario D is the established policy basis of Egyptian water utilization policy toward Ethiopia.

And last, that these rules have several confusing and vague Articles that, according to Egypt, put Ethiopia on a wavering legal position grappling with their "true meanings" which easily would shift the "burden of making prior cares" to Ethiopians (Abate 1990). However, these international rules never explicitly provide for non-use of upper riparian states to utilize international rivers for various purposes on their domestic territories except that they make it a duty for the state to assure that it does not negatively affect the lower riparian states. Let us see how the instruments of "Equitable Utilization Doctrine" of international law appear to be meshed up and confused with the principle of "no harm" (Alebachew 2011).

Relevant factors: "Relevant factors which are to be considered, include, but are not limited to: (a) the geography of the basin, including in particular the extent of the drainage area in the territory of each basin State; (b) the hydrology of the basin, including in particular the contribution of water by each basin State; (c) the climate affecting the basin; (d) the past utilization of the water of the basin, including in particular existing utilization; (e) the economic and social needs for each State; (f) the population dependent on the waters of the basin in each basin State; (g) the comparable costs of alternative means of satisfying the economic and social need of each basin State; (h) the availability of other resources; (i) the avoidance of unnecessary waste in the utilization of the waters of the basin; (j) the practicability of compensation to one or more of the co-basin States as a means of adjusting conflicts among uses; and, (k) the degree to which the needs of a basin State may be satisfied, without causing substantial injury to a co-basin State."

If these are the conditions of equitable utilization of international waters, what does "no harm" principle mean, therefore? Who, and how, can measure the incidence of "harm" on a riparian state by the actions of another upper riparian state, be it diversion or otherwise?

What if hundred or one thousand electric lumps lose their bright color powers across the streets of Cairo following the construction of the Grand Abbay Dam by Ethiopia? Is it "harm" that could be cited as enough to justify Ethiopia's refrain from constructing the dam? These questions have been common wrangling points among the diplomats of concerned states so far. According to Mohammed Amr, Ethiopia has supported the idea of qualifying the principle "no harm" to be more operational adding such qualifications like "no significant or appreciable harm" (Amr 2002). Egypt was opposed to the qualification proposal preferring the term to remain vague. However, Mohammed himself admits that the "no significant harm" principle is more acceptable than the "no harm" among many diplomats and legal scholars. Ethiopia, on its part, advances these international rules and conventions for specifically three major reasons: first, distinguishing technicality from fundamental international principles of cooperation is one of the most important challenges of Ethiopia to the Egyptian stance on "non-diversion" principles (Whiteman 1964). According to Ethiopians, diversion may induce not only negative results but also positive improvements of the overall utilization of Nile river waters. Ethiopia argues that any water utilization on its territory could be mutually agreed through bilateral and multilateral cooperation and consultation over its technical effects among professionals in the field but what matters most is the political will (Whiteman 1964).

For Ethiopia, the principles of "no-diversion, equitable utilization of the Nile Waters and no-harm" arguments could be reconciled only through close interstate mutual understanding. Ethiopia capitalizes on the "equitable utilization" doctrine and prefers no-diversion and no-harm principles to being tabled to interstate technical verifications. In the absence of formal bilateral or multilateral treaties over the use of Nile River, the remaining legal option for Ethiopia appears to stick to Article 2 of the Helsinki Convention (Arsano 2003) on the Protection and Use of Transboundary Watercourses and International Lakes 1992. It states that the State Parties to the Convention shall take all appropriate measures to ensure that the transboundary waters are used in a "reasonable and equitable" way. Moreover, this principle originated as a middle position of reasonableness between the two extreme principles: the absolute territorial sovereignty assertion of upstream states, and absolute territorial integrity claims of downstream states (Utton 1996). Many terms have been used for this principle, such as "equitable apportionment" or "equitable participation" where most propositions have proved to be to the chagrin of Egypt; That the rules advance more the principles of "equitable share and no harm" is a well-established principle of international law that each riparian state is entitled, within its territory, to an equitable share of the water from an international watercourse. The principle of equitability has been recognized by international doctrine and several bilateral and multilateral treaties, including the UN Convention 1997 after many efforts to make the term much clearer to enable interstate cooperation to come true. Ethiopia expresses strong commitments to the principle of "no-harm" by basing its arguments on three fundamental policy and technical justifications (Policies and Strategies of the EFDR Government 2002): Ethiopia does not need the Nile Water for agricultural purposes except a few amounts as little as 6 billion L^3 which is extremely negligible for Egypt. Egypt wastes more than two times of this amount of water through apparently careless management of evaporation. The significant demand of Ethiopia for water is for the development of hydroelectric power. The used water will continue to flow to lower riparian states avoiding the confusion of diversion. Such experiences abound in many interstate conflicts over the use of international rivers for same goal (Caponera 1992). Spain leveled a similar case in 1890 on France tabled for adjudication through an Arbitration Tribunal. It was, however, finally proven that the Hydroelectric Power Plant never diverted or reduced the amount of river waters, nor adversely affected the environment whatsoever.

Ethiopians argue that utilization of the Nile water for HydroElectric power through the Grand Abbay Dam is much economical and effective, incurring no perceivable wastage through evaporation because of its high altitude. As compared to Egypt and Sudan, which waste as much water as 12 billion L^3 through evaporation, and as high as 32 billion L^3 in evaporate-transpiration, Ethiopia's loss is less than a few millions, which has no a perceivable consequence on the former. Egypt's huge loss of considerable water through evaporation could only be remedied by such projects like the Renaissance Dam, which will substantially address the problem (Dingamo Reporter 2011); Ethiopia also strongly believes that its development activities over the Nile River will have a positive effect on environment which ultimately will benefit Egypt and Sudan for three reasons: one, it helps to rehabilitate the largely deforested basin due to recurrent droughts in the three major states; two, it encourages watershed development in which the rehabilitated and cultivated portions around the Grand Abbay Dam Project and as far as Sudan will store an unprecedented amount of ground water to support and sustain the overall Nile surface water and facilitate a healthy hydraulic process in the region; three, it significantly arrests erosion reducing the considerable silt, negatively affecting and wasting water in Egypt and Sudan. According to J.V. Sutcliffe and J.B.C. Lasenby, the amount of loss in water and agricultural productivity in Egypt and Sudan only in 1986–88 due destruction caused by heavy flooding equals the total amount of the water wasted through evaporation in an average of 5-year period (Sutcliffe and Lazenby 1990). Zewude Abate notes that if Egypt and Sudan are committed to fully utilizing the Nile Waters for their development needs, wastages of water, destructions and harms both from droughts and floods, and seasonal fluctuations should be mitigated significantly and ultimately controlled (Abate 1990). The key instrument for this goal to achieve, however, is to be found in Ethiopia.

That these duties are simpler to meet makes them highly preferable by Ethiopia. For Ethiopia, there are several experiences where interstate treaties could prevent potential conflicts simply because they rule out detailed mechanisms of peaceful settlement of disputes in advance of their occurrence (Hayton 1967).

References

Abate, Z. (1990). History of the Nile flows. In P. P. Howell & J. A. Allan (Eds.), *The Nile: resources evaluation, resource management, hydro-politics and legal issues*. Prepared for Publication for a Conference held in London at the Royal Geographic Society and School of Oriental and African Studies, University of London.

Addis Zemen Daily. (2011). *The Editorial at Addis Zemen Daily* (Vol. 64). April 4, 2011.

Alebachew, H. (2011). International legal perspectives on the utilization of trans-boundary rivers: The case of the Ethiopian renaissance (Nile) dam. Mekelle University, College of Law and Governance, Ethiopia, Presented to the Ninth IUCN Colloquium, North West University of South Africa, Eastern Cape, July 2011.

Amr, M. S. (2002). Diversion of international watercourses under international law. In (Eds.), *African yearbook of international law* (Vol. 10). Leiden/Boston: Martinus Niijoff Publishers.

Arsano, Y. (1990). Arab Alliance and Ethiopia's reaction to the Red Sea question. *Paper Presented at Peace and Development in Ethiopia Symposium, Organized by the Main Department of the Higher Education of Ethiopia.*

Arsano, Y. (1999). Predicaments of cooperation in the Nile Basin. In K. Fukui et al. (Eds.) (1997) *Ethiopia in Border Perspective, the Proceedings of the 13th International Conference of the Ethiopian Studies*. Kyoto, Japan.

Arsano, Y. (2003). Trans boundary waters: prospects for peaceful co-existence in the Horn of Africa. In T. Olika, Y. Arsano, & O. Aadland (Eds.), *in Topics Contemporary Political Development in Ethiopia. Proceedings of the Launching Workshop of the Department of Political Science and International Relations(DPSIR) and the Norwegian Institute for Human Rights(NIHR)*. Addis Ababa: Addis Ababa University.

Arsano, Y. (2011a). The ETRE, Yegna Guday (Our affaire) weekly forum, May 29, 2011.

Arsano, Y. (2011b). So remarked on Studio Forum called 'our affaires' held on May 24th by the Ethiopian National Television specially prepared on the issues Renaissance Dam.

Ayele, N. (2005). Who was civilized and who was savage? In *Poulos Melikias and Get chew Metaferias, the Battle of Adwa*. New York: Algora Publishing.

Bartnisky, A., & Neicheko, Y. M. (1981). *Ethiopia's history.*
Berber, F. (1959). *Rivers in international law.* London.
Caponera, D. (1992). *Principles of water law and administration.* London: National and International, Queens Publications.
Dingamo, A. (2010). Reporter for interviews with Asfaw Dingamo, Minister of Water Development from 2007 to 2010.
Dingamo, A. Reporter. (2011). Interviews with Afsaw Dingamo over Ethio-Egyptian policy cleavages over the utilization of River Nile. Volume 16, Number 13.
Egypt and Sudan agreement. (1959). Egypt and Sudan signed this agreement in 1959 in Cairo as major lower riparian states which provides for the Full Utilization of the Nile Waters with 55 and 18 billion cubic liters of water share between them respectively.
Evans, E. (1990). History of the Nile flows. In P. P. Howell & J. A. Allan (Eds.), *The Nile: resources evaluation, resource management, hydro-politics and legal issues.* Prepared for Publication for a Conference held in London at the Royal Geographic Society and School of Oriental and African Studies, University of London.
Foreign Policy and National Security Policies' Document. (2011). Foreign policy and national security policies' Document page 14.
Gibb, A., & Partners. (1978). "Hunting Technical Services" Blue Nile Waters Study, Volume 1, Main Report.
Guariso, G., et al. (1987). Implications of Ethiopian water development for Egypt and Sudan. *Water Resources Development, 3.*
Hayton, R. (1967). The Plata Basin. In A. Garretson et al. (Eds.), *The law of international drainage basins.*
Helsinki Rules. (1966). Helsinki rules of international water courses and the 1997 United Nations convention on non-navigational uses of international watercourses.
Kendie, D. (1999). Egypt and the hydro-politics of the Blue Nile River. *Northeast African Studies, 6*(1–2), 141–169 (Published by Michigan State University Press).
Knott, D. G., & Hewett, R. M. G. (1990). History of the Nile flows. In P. P. Howell & J. A. Allan (Eds.), *The Nile: Resources evaluation, resource management, hydro-politics and legal issues.* Prepared for Publication for a Conference held in London at the Royal Geographic Society and School of Oriental and African Studies, University of London.
Mc-Caffrey, S. C. (1996). The harmon doctrine one hundred years later: Buried, not praised. *Natural Resources Journal, 36.*
McCann, J. (1981). Ethiopia, Britain and Negotiations for the Lake Tana Dam 1922–1935. *International Journal of African Historical Studies.* U.S. Department of the Interior, Land and Water Resources of the Blue Nile Basin: Ethiopia 17 Vols. Washington, DC: Government Printing Office, 1964.
Oppenheim, L. (1955). *International Law* (8th ed., Vol. 1). Edited by H. Lauterpacht.
Policies and Strategies of the EFDR Government. (2002). Ministry of Information, Foreign Affairs and National Security, Policies and Strategies of the EFDR Government: Ministry of Information, Department of Press and Audiovisual, Addis Ababa.
Reda, G. (2011). A spokesperson for the foreign Ministry of the Ethiopian government said it the ETRE, Yegna Guday (Our affaire) weekly forum, May 29, 2011.
Reta, Z. (2000). *The Eritrean Affair: From 1941 to 1962.* Addis Ababa: Shamma Books Publishers.
Sutcliffe, J. V., & Lazenby, J. B. C. (1990). History of the Nile flows. In P. P. Howell & J. A. Allan (Eds.), *The Nile: Resources evaluation, resource management, hydro-politics and legal issues.* Prepared for Publication for a Conference held in London at the Royal Geographic Society and School of Oriental and African Studies, University of London.
The Ethiopian Renaissance Dam. (2011). The Ethiopian Renaissance Dam' is the official name given to the new Blue Nile Dam by the 66 Regular Session of the Council of Ministers of the Federal Government in May 2, 2011.
The newly launched Nile Basin. (2009). The newly launched Nile Basin has made this doctrine the basis of its entire goals and policies on its Establishing Treaty signed in 2009.

Tvedt, T. (1992). The management of water and irrigation: the Blue Nile. In M. Doornbos et al. (Eds.), *Beyond conflict in the Horn of Africa*. London: James Curry Limited.

Utton, A. (1996). Which rule should prevail in international water disputes: That of reasonableness or that of no harm? *Natural Resources Journal, 36*.

Weekly (2011a). Reporter Weekly, April 28, 2011, Vol. 16, No. 227.

Weekly (2011b). Reporter Weekly, March 23, Vol. 16, No. 222.

Whiteman, M. M. (1964). *Digest of international law*. Washington, DC: Department of State Publication.

Yenus, S. (2011). The Ethiopian National Television transmitted the news on its May 8th Evening News Desk coating General Samora Yenus, the Commander-in-chief of the armed forces, as the official source.

Zenawi, M. (2007). Prime Minister Meles Zenawi accused of the former Libyan leader, Colonel Moaamour Gaddafi in an interview he held on March 21, 2007 with domestic private and public journalists of conspiring to shift the African Union Headquarters from Addis Ababa to Tripoli.

Zenawi, M. (2011a). Prime Minister Meles Zenawi gave contradictory answers for Egyptian National Television on October 10, 2011 where he underpinned the existence of cordial relations with both states but accused Egypt for conspiring against Ethiopia's national interests in an interview on April 2, 2011 for domestic Television and Radio journalists while he consistently assured relations of friendship with Sudan. However, his official policy statements at a state visit on September 16, 2011 in Cairo underscored general friendship and cooperation with either Egypt or Sudan.

Zenawi, M. (2011b). The Prime Minister of the Federal Government of Ethiopia informed the Ethiopian people in a televised speech detailing size and budget estimate of the Dam on April 4/2011 after he officially inaugurated the beginning of its construction.

Chapter 10
Efficiency Use of Irrigation Water

Abstract This chapter provides a brief overview of the efficiency use of irrigation water in Egypt, background of efficiency use of irrigation water in Egypt, scope of the chapter, political development model of delta, ecological definition, social definition, political definition, the three stages of development of a delta, the three stages of development of a delta in Egypt and mathematical model of efficiency use of irrigation water in Egypt.

Keywords Scope of the chapter · Political development model of delta · Ecological definition · Social definition · Political definition · The three stages of development of a delta · The three stages of development of a delta in Egypt · The role of commodities and efficiency in agricultural growth in Egypt and mathematical model of efficiency use of irrigation water in Egypt

The Nile River in Egypt is the primary source of water. It is used for strategically important water uses such as drinking, fishing, industrial use, livestock and irrigation. The water in the Nile in Egypt is intricately managed through an extensive system of dams, barrages, and canals. The water from the Nile is conveyed to the users through a vast network of canals. Wastewater and agricultural drainage water from these uses are collected by drains and are often returned to the Nile River as inflows.

Any disruption or impairment to the Nile River from natural or anthropogenic threats can potentially have far reaching economic and social implications. Approximately 99% of the population of Egypt (total population is approximately 85.783 million) lives within the Nile Valley and delta, which constitutes less than 4% of Egypt's total area. The area of the Nile Delta itself is about 25,000 km^2 and has approximately 35 million inhabitants (Abdel-Gawad and Khalil 2003).

© Springer International Publishing AG 2017
Y.M. Hamada, *The Grand Ethiopian Renaissance Dam, its Impact on Egyptian Agriculture and the Potential for Alleviating Water Scarcity*,
Environment & Policy 55, DOI 10.1007/978-3-319-54439-7_10

Background

'Conflict' is a word that has a variety of definitions covering a range of experiences and issues. Perhaps this is one obvious reason why agreement on resolving conflicts can be elusive. Root causes of conflict, such as poverty or disease, remain largely untreated. Apart from disagreements on definitions, problems also arise over statistical evidence, as witnessed in the Cold War when people argued about the number of warheads on rockets or the types of planes and ships that **NATO** and the Warsaw Pact could use. The costs of weapons systems cause other differing perspectives, as well as the numbers involved in conflict—civilians, military and the numbers of dead and injured. Ultimately, people may ask 'When is a conflict not a conflict?' Personalities make or break conflicts and all types of conflict behavior issue first from their leaders' minds: the suspicions and paranoiacs of dictators such as Stalin, Pol Pot and Kim Il Sung, or fears about the resurgence of global anti-Semitism and fascism. Peace overtures in areas where conflicts appear intractable, such as Kashmir and Palestine, have foundered and others, such as Northern Ireland, Guatemala and Sudan, have shakily succeeded only after decades of hesitant peacemaking (Thackrah 2009).

Often these interminable problems arise as a result of minority issues, especially the overlap between split minorities and disputed frontiers, which have spawned disorder most notably in former Yugoslavia and the Middle East. Another cause of conflict is control of resources such as coal, oil and water. This was true throughout the Cold War and still is in the period of the New World Order, which again as a term in international relations is hotly contested. One area of potential conflict that has become less important since the 1990s is outer space, because of the Soviet collapse and the enormous costs involved. Weaponry has moved on to the general use of missiles and rockets (born out of the V1 and V2 missiles of the Second World War) and more recently to weapons of mass destruction. Politicians and governments may get more kudos from military conflict than from ordinary domestic issues. People and organizations rarely keep to treaties and alliances, obligations and commitments. In some cases they are stymied by issues such as international terrorism, Islamic fundamentalism, rogue states and environmental factors, such as control of resources—oil, water, natural gas, and plutonium—on land and under the sea. Maritime boundaries can be just as contentious as land frontiers. Wilderness politics has become more noticeable in recent years—for example, in the Arctic, Antarctic and Siberia. Dealing with pandemics such as AIDS, SARS and bird flu can cause rivalries to come to the fore (Thackrah 2009).

In 1945, superpowers were created with the birth of the nuclear age, and great powers and ordinary powers have to live side by side. One superpower has fragmented—the Soviet Union—but others wish to take their place. China is the logical heir to the title, but this could take some time. Many tensions have existed within and around its borders and have the potential to do so again. Globalization, it is argued, has made the world a safer place but many would argue that states are controlled increasingly by international organizations or multinational companies. The global village is a far cry from a world carved up by colonial powers or ideologically motivated ones. Yet many in the **UN**, **EU**, **OAU** and **OAS** argue that such organizations could do more to maintain world peace. The world has definitely moved on from being Eurocentric to being polycentric. New states have proliferated and many see themselves as neutral or perhaps non-aligned with the increasingly free diplomacy since the end of the Cold War. Legal issues have to prevail, to formally end conflicts. Some would ask whether military conflict is just, or has it been superseded by diplomatic, economic and political conflict? The shadows of fascism and anti-Semitism, the persecution of religious groups, ethnic rivalries and genocide—which have clouded so many conflicts throughout history—all still exist. The capacity for great evil lies undiminished. The great expanse of military conflict over the past six decades shows clearly the dangers of excessive patriotism and nationalism. The suspension of human rights, even in democratic countries, on its own does not lead to the ending of conflict. The totalitarian temptation is always there (Thackrah 2009).

Scope of the Chapter

On the March 5, 1979 the spacecraft Voyager 1 reached the Jupiter system and began to transmit a steady stream of spectacular images back to Earth. The one-tone vehicle had been launched eighteen months earlier from the NASA Kennedy Space Center at Cape Canaveral, Florida, USA. Voyager's trajectory past Jupiter was designed so that the planet's enormous gravitational field could be used to alter the spacecraft's heading and accelerate it on toward Saturn a maneuver called a "gravity-assist swing-by". The Jupiter swing-by added some 60,000 km/hour to the speed of the spacecraft, which made its rendezvous with Saturn on 12th November 1980. Its twin, the spacecraft Voyager 2, followed an even more complex path, using three gravity-assist swing-byes to complete its Grand Tour flight to Neptune via Jupiter, Saturn, and Uranus. Voyager 2 swung past Neptune on August 25, 1989, a full twelve years after its launch from Cape Canaveral on 20th August 1977. On 23th April 1990, seven months after Voyager 2 passed Neptune and headed out into interstellar space, the small town of Nyngan in central New South Wales, Australia, was inundated by the Bogan River. After a week of desperate effort by the residents, during which time some 250,000 sandbags were used to build up those sections of the permanent levees that had low freeboard, one of the sandbag levees breached and the floodwaters poured into the town. The town's 2300 residents were evacuated by helicopter and some $47 million worth of damage was inflicted in the urban area. According to the New South Wales Department of Water Resources (**DWR**) the "April 1990 flood devastated Nyngan: it disrupted the social, financial and community foundations of the town; it imposed tremendous financial loss and psychological despair on the residents". This was not the first Bogan River flood to cause problems in the area the historical record indicates that flood waters have reached the Nyngan Post Office a number of times since the township was first "submerged" in July 1875.

From the earliest times policies designed to improve the management of water resources have had counterproductive outcomes (Tainter 1988). For example, Jacobsen and Adams (1958) describe the likely contribution of irrigation developments, via accumulations of salt and silt, to the breakup of ancient Mesopotamian civilizations. Modern approaches to river control still produce unanticipated problems it is not unusual for flood-risk reduction policies to result in increased actual risk (Smith 2000). For example, proponents of the "levees-only policy" reject the use of other structural or natural measures (such as cut-offs or floodplain storage in tributaries and wetlands) and social measures (such as reduced floodplain occupancy) despite evidence that levees often fail to work as anticipated and can increase the risk of flood-induced damage. The great Mississippi floods of 1927 and 1993 provide powerful examples (Changnon 2000). The debate over the value of levees continues to this day (Hallowell and Jehl in the New York Times, April–May, 2001).

Why, in an age when scientists and engineers can develop effective spacecraft navigation policies, do flood-mitigation policies still produce unintended and unwanted outcomes? Our answer is that, from the dynamical point-of-view, it is much easier to design and implement good spacecraft-navigation policies than to design and implement good flood-risk mitigation policies. While the Voyager scientists and engineers faced daunting technical challenges, they knew the simple dynamical laws (established in the seventeenth century by Isaac Newton) that would govern the trajectories of their spacecraft. In contrast, those who established the Nyngan flood-risk mitigation policies were forced to deal with a system whose dynamical laws are much more difficult to discover and understand. A river system in flood is driven by complex and contingent interactions between weather, topography, turbulent flows, sediments and soils. When such a system is tightly coupled with a human social system, which has behavior that is even more complex and contingent than that of the natural system, it becomes very difficult to design policies that work as intended (Newell and Wasson 2002).

Management studies carried out over the last 40 years, under the general heading of "System Dynamics", have made it clear that policy makers need to understand the basic dynamics of feedback systems if they want to avoid surprise and to identify leverage points for system improvement (Forrester 1961, 1969; Senge 1990; Sweeney & Sterman 2000). In this paper we use one of the tools developed by the System Dynamics community, causal-loop diagrams, to structure a discussion of some possible reasons why the levees-only policy has unintended outcomes and why it persists despite its limitations. Our analysis is intended to provide a glimpse of the way that internally-generated forces dominate the behavior of complex systems. We contend that policy making, as a fundamental component of society's adaptive process, needs to have an experimental focus to compensate for the weakness of forecasting as a management strategy in coupled natural/social systems. In such situations the history of system behavior is an essential source of learning because it contains an invaluable array of completed, long-term "experiments" (Newell and Wasson 2002).

Political Development Model of Delta

Through three contrasted deltas, agricultural water is not inevitably a good guiding thread for the history of a fitted out region, nor water is a good guiding thread for the history of a wet region. The omnipresence of water does not guarantee indeed the continuity of stakes, the continuity of infrastructures or still the continuity of economic functions of water. As a result, we suggest the absence of any single pattern of historic development in deltas. The delta ecology displays an internal organization which can be characterized as a "sequential mosaic"; then we propose a social definition and we sketch a more political definition based on stakeholders in terms of power and social representations which underpin and justify the decisions of the dominant groups.

Ecological Definition

The delta is a deposit of clay, silt and sand formed at the mouth of a river where the stream loses velocity and drops part of its sediment load (triple idea of earth (earth, heaven and humanity), horn and islands (ground and water). The successive bed shifting (influence) results in an intricacy of alluvial "lobes" or "sheets" where more or less salted and coarse alluviums mix (Dacharry 2000).

Superficially soils mirror such interweaving and gradients and the spatial distribution is complicated with former salted water wedges far inland and former sulfate acid soils (Doras). Former levees and sandy bars at old mouths built a kind of more or less apparent, raised grid. This widely crossed substratum is itself affected by a uneven sensibility to regular or occasional, deep or superficial flood and, in more or less effective drainage. In dry season, this wet environment becomes a huge barren area, sometimes with poor vegetation and difficult human settlement. Those multiple factors, before any human activity, results in a large variety of wet environment. One foot in height make difference between two areas akin, to the point where internal variability is more important than that distinguishing the delta itself with the river valley upstream or surrounding regions. Obviously, the image of the Nile surrounded with the desert is misleading because it is specific. As a result, the same minimal height modifies human activity, as shown by staggered crop systems depending on the probability to be touched by the flood in Bangladesh. However, symmetries and sequences organize a delta. If they do not inevitably give a consistent distribution, they make easier the implementation of a model from which abnormalities can be underlined and discussed (Mollard 2001).

So each delta can be divided in many parts. We simplified them with a succession of three zones. Even with such a simplification, delta landforms are very variable: Coastal Delta is the permanently wet zone with brackish waters in dry season and fresh water in wet season; the Middle Delta is a relatively flat zone where wide, low flood spread; the Higher Delta is the maze of levees with more or less temporary ponds. It includes the flooding plain with regular, deep floods (Mollard 2001).

Social Definition

A delta is a territorial stake and a resource stake. History shows two opposite types of deltas: those with an ancient organization and those with late coming ones, these latter showing that, according to the society, delta wet environment is not a priority. For the Nile and the Chinese deltas, history is unclear and begins with the writing and the administration, making very speculative the earlier events. On the other hand, Camargue, Southeast Asian deltas, Medjerda in Tunisia, those of the New World, etc. have been reclaimed in the last centuries, and some African deltas remain idle (Lacoste). Resources in wood, fishing and salt have been exploited at a particular moment. Such time scheduling also exists for hydraulic reclamation with the possibilities of silt-based soil fertility, control of pests and weeds by flood, irrigation, river communication, even for military purposes, source of energy and drinking water supply to cities. During time, the economic functions of water diverged, interwove, changed streams, and even disappeared. However, rights, institutions and sometimes equipment are of a particular period deposited and headed following events. Sometimes, sparse population with poor previous rights on such a huge territory has been rushed away by invaders. As a result, the concerns of a delta are, on one hand, bound to the history of surrounding regions and, on the other hand, to political systems. A delta is suitable for any kind of despotism, and not only in Eastern countries as soon as there is a significant surplus to extort and because water management suits large public works. Wittfogel (1956) is not wrong, but it is not possible to generalize his view.

Political Definition

Delta environment is convenient to large-scale reclamation for two reasons: institutional because of badly asserted rights and extensive resource management, and geographic because of an extensive, flat and aquatic area where civil engineering and labor mobilization are often previously required. Individual reclamation as raised-bed plots ("hortillonnages"), poldered set of fields and levee openings (as in the Niger delta today) are possible in a more progressive way. However, the extent of disasters or the scope of economic stakes contributes to doing a preserve of the administration whose answer is the hugeness of civil engineering. It is not possible to understand a delta without this double reference to good and evil which are the two faces that administration wields. So it justifies the work supervision and control of water by fighting against disasters and by trying to benefit from agricultural abundance. The strength of such a double-faced mental scheme associated to water is that it is accepted without discussion and without perception of ambiguities. It is very convenient for the administration, and current social representation is here comparable to an ideology or a religion as in antique Egypt (Mollard 2001).

The Three Stages of Development of a Delta

Before agricultural reclamations and up to roads and railway building, the history of a delta began by easily navigable ways and relations with nearby and distant regions by sea. In this respect, the initial pattern of the three deltas was different. As an interface between sea and mainland a delta was at first convenient to the business. As Huang (1990) writes "It was the distinctive transport network of the delta and its low cost water links to the middle and the upper Yangzi regions and the southern seacoast that permitted a higher degree of evolutionary commercialization of the peasant economy than in a region such as North China". The internal or close urban network and the economic environment, which is the array of regions and resources outlined intensity and direction of commercial roads. Orientation of canals and exchanges entailed important consequences for long and the situation of the delta within a national territory mattered. On the other hand, difficulties always existed to create a port in a delta because of few possibilities of mooring for boats and because of the sand bar in the mouth (Rosette branch for the Nile, Chao Phraya, silting of Aigues-Mortes when Marseilles did not belong to the king of France).

Another precedent was the preliminary appropriation of lands. Unlike wet valleys and inland swamps, an extensive delta did not lead to massive ownership unless a strong social group was involved. It has been the case of cattle breeders, which often belonged to major political, warrior organized groups. Indeed, an actual settlement often required preliminary works beyond the banks. Without preliminary rights, administration was able not only to be necessary but it sometimes acquired a monopoly making the farmers depending on the access to the water and to the land. Second phase was that of agricultural reclamation without considering the delta as a whole. Every part of the mosaic has been reclaimed according to its resources and its own environment. The State intervened, but, even in Egypt, it only improved a peasant or private system which was an adaptation to local conditions. The environmental "artificialisation" might be technologically sophisticated, but there was no will of systematic exploitation or voluntarism which distinguished the third phase. This absence of global attitude was not a shortcoming as Occidental engineers later claimed (Mollard 2001).

The Three Stages of Development of a Delta in Egypt

In the three deltas, crops, techniques and chronology of events were different. The reclamation of the Nile delta dates back several millenniums. Geographic continuity and homogenization between the valley and the delta were remarkable. The flood was spread owing to hoes or checked-layout bonded fields with a collective intake upstream and a collective outlet for complete draining. Challenge was to regulate a fluctuating phenomenon from one year to the other one and to protect land from inconvenient inundations. Maybe, the height of the dike was originally lower to be enough to only trap waters and silt. Human settlements required elevations with local embankment (Which remains an important activity in unprotected deltas such as in Thailand). The progressive extra height of the dike and the digging of long feeder canal for remote new lands doubtless increased water control and regularity: no reservoirs as in Sri Lanka, no irrigation as it was the case later (except for some gardens). It was not an irrigated or submerged farming, but decree farming. Dryness was a problem although wheat more than sorghum was the main crop. Fallow was useless, labor needs were reduced and productivity widely exceeded the population needs (Mollard 2001).

The Pharaohs dug several big canals which connected the arms of the Nile, as well as canals for flood spreading in inter fluvial areas (interfluve). After the farmers or the noble persons, the State prematurely substituted for the peasant communities. In fact, recent reclaimed deltas show that every kind of institutional reclamation was possible: private entrepreneurship of feudal landowners, peasant reclamation or state projects. On the other hand, conflicts and especially a very strong productivity and surplus extortion paved the way to strengthen no-producer classes and administrative agencies to justify such classes. Religion as ideology, judges of waters stemming from the priest class and scribes to collect taxes, rule irrigation and raise dikes were at the origin of a highly organized civilization among the first in the world. During this stage, no vision of the delta as a whole prevailed. Lower Egypt was distinguished for political reasons (independent kingdoms and capitals before the reunification, strategic to the sea) more than its own resources: salt, papyrus, etc. were complementary in the same way that upstream production (Mollard 2001) (meadows, Nubian cattle production, etc.).

Under the Arabic and especially Turkish empires, the delta experienced a strong regression with the abandonment of numerous canals and their blocking with sand from the desert and silting. Jaubert of Passa carried out a fine institutional analysis. For him, the Arabs had already destroyed the pillar of the religion; with the Turks, the subjection of Egypt in the suburb of the Empire according to a social code of exploitation would be the major cause of the crisis. The voluntarism of Mohamed-Ali, maybe more modern than rooted in the history of the Egyptian despotism, led the country to unequalled performances (even though the fellah is more than ever exploited).

The Role of Commodities and Efficiency in Agricultural Growth in Egypt

Agriculture is an issue as a local food source, for international trade, for balance of payments, land use and water use and as a basic product for food and fiber manufacturing. Hence every aspect of the economic structure of a country relates to agriculture. Banking, transportation, tax and tariff structure, subsidies, local and international markets and health are all part of the agricultural system of a country. Egypt is no exception. Agriculture is the country's largest employer accounting for about 28% of the labor force and 13.9% of the GDP in 2005.

Agricultural production is intensive and yields are high, but only 3% of land is arable. In spite of land reclamation, the area under cultivation remains about constant because of urban and industrial expansion. With no land expansion and population growth, Egypt will remain one of the world's largest food importers. Some 95% of local production is consumed domestically despite the increased emphasis on cash crops for export. The growing population of Egypt needs employment and rapid agricultural growth results in the quickest employment growth. Agricultural growth of 5% can be achieved through increased productivity, efficiency and competitiveness and could result in about 500,000 jobs. This profound impact on employment growth requires major improvements in policy matters; however, policy constraints continue to prevent Egypt from fully realizing the comparative advantage it has in most commodity systems. Should Egypt focus on urban or agricultural growth? It is rural and agricultural growth, not urban or industrial growth that reduces poverty and increases demand for labor. These findings are consistent with the fact that agriculture drives the demand for labor through its demand for goods and services produced in the large, labor-intensive, rural non-farm sector.

GDP growth depends largely on the ability to expand production in the tradable sectors, while employment growth depends largely on increases in (domestic) demand for non-tradable. When agriculture grows rapidly, demand for labor grows rapidly; when urban tradable grow rapidly, GDP grows rapidly. A structure weighted toward agriculture is weighted toward benefits to labor; a structure weighted toward urban tradable items is weighted towards fast growth in GDP. Urban tradable grow primarily through increase in the capital stock. Agriculture tradable grow largely through technological change. Treating "rural" and "urban" poverty as somehow separate and in competition with each other for resources is not only a conceptual mistake, but a remarkably short-sighted view of the problem. In fact, successful rural development generally stimulates and supports urban development. Conversely, urban growth is a powerful stimulus to food production, especially by small farmers. The projected expansion of the urban population in Asia and Africa, from 1.7 to 3.4 billion over a period of only 30 years, and the reduced level of available resources, stress the need for a more imaginative but pragmatic response.

In this discussion, employment increase and poverty reduction are used interchangeably because increased employment is the means by which growth reduces poverty. The World Bank has made frequent use of an average relationship between overall growth and poverty reduction. On average a 2.1% decrease in poverty levels is associated with a 1% increase in the economic growth rate.

An analytical model for Egypt, finds that, compared with pro-manufacturing policies, pro-agricultural policies have a more positive impact on household welfare in general and the poor in particular. It also shows that improved market access for agricultural exporters and reduced transaction costs in foreign trade reinforced the effect of generally positive policies toward agriculture. Raising farm incomes—which can be achieved through technological change and globalization—increases the demand for output and hence for labor in the massive rural non-farm sector that is home to the bulk of the poor and underemployed.

That is why agricultural growth is not only important to growth in national income, but also absolutely vital to growth in employment and reduction of poverty. To discuss agricultural growth in Egypt discuss the Nile, Fig. 10.1.

This chapter describes the commodity composition for a high growth rate for Agriculture. Farmers make decisions about resource allocation among specific

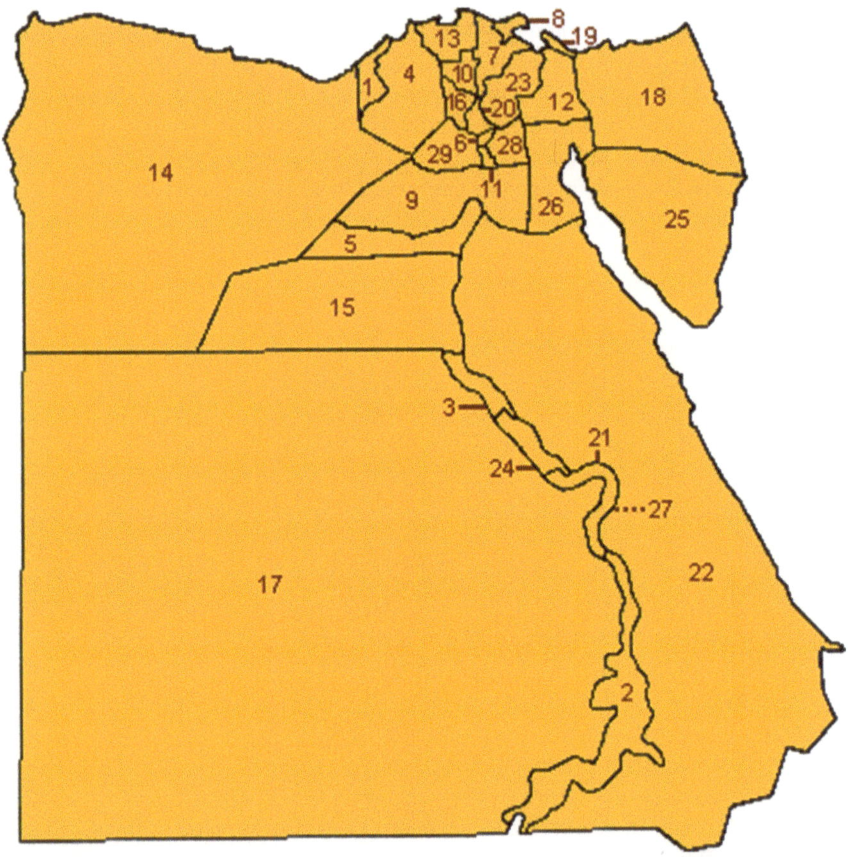

Outside the Valley	Upper Egypt	Middle Egypt	Lower Egypt	
17. New Valley	3. Assuit	9. Giza	19. Port Said	1. Alexandria
14. Matruh	24. Sohag	5. Beni Suef	20. Sharkia	10. Gharbia
25. South Sinai	21. Qena	29. Fayum	23. Damietta	16. Menoufia
18. North Sinai	27Luxor	15. Mania	26. Suez	12. Ismailia
27. Noubaria	2. Aswan		4. Behera	13. Kafr-El Sheikh
			11. Cairo	6. Qalyoubia
				7. Dakahlia

Fig. 10.1 Map of the Arab Republic of Egypt-Locations of the governorates in the zones. *Source* Hamada (2014)

commodities. Much of investment, institution building, and policy are also commodity specific. For all commodity groups, two inputs, water and fertilizer, are particularly important to achieve strong growth rates. Fertilizer is already used at very high rates in Egypt. Increasing the efficiency of fertilizer use will favor higher farm incomes as well as being environmentally sound. Efficiency of water use is also critical to achieve strong growth rates. Egypt has highly productive agricultural resources that are reflected in high crop yields. Such conditions are normally highly responsive to the new opportunities offered by technological advance and the forces of globalization.

There is an implicit assumption that new lands will be allocated optimally and will be comparable in quality to past additions. It also is assumed that sufficient increase in water use efficiency will occur to meet the rapidly growing urban needs without productive loss in agriculture. These are all heroic assumptions. For them to become reality will require continuing attention to water use policy and the making of hard decisions about where new water will be allocated. In this context, difficult conflicts between the best allocation for employment and national income growth and broader national objectives will have to be resolved. Not Achieving the Comparative Advantage.

Thus, the solution to Egypt's employment problem can be met only with major efforts in agricultural employment as well as accelerated growth of the urban, tradable goods and services sector. It is not a matter of one or the other; it is a matter of the total of both sectors. The importance of the task is challenging and enormous. A recent statement from the Office of the Minister of Commerce and Industry of Egypt in July of 2007 estimated that one million new jobs are needed in Egypt each year.

Simulate Alleviate Water Scarcity Crisis System (AWSCS) Can Be Written as

Simulate

$$\sum_{a=1}^{A}\sum_{l=1}^{L}\sum_{l=1}^{L}\sum_{e=1}^{E}\sum_{v=1}^{V}\sum_{i=1}^{I}\sum_{a=1}^{A}\sum_{t=1}^{T}\sum_{e=1}^{E}\sum_{w=1}^{W}\sum_{a=1}^{A}\sum_{t=1}^{T}\sum_{e=1}^{E}\sum_{r=1}^{R}\sum_{s=1}^{S}\sum_{c=1}^{C}\sum_{a=1}^{A}\sum_{r=1}^{R}\sum_{c=1}^{C}\sum_{i=1}^{I}\sum_{t=1}^{T}\sum_{y=1}^{Y}\sum_{c=1}^{C}\sum_{r=1}^{R}\sum_{i=1}^{I}\sum_{s=1}^{S}\sum_{i=1}^{I}\sum_{s=1}^{S}\sum_{s=1}^{S}\sum_{y=1}^{Y}\sum_{s=1}^{S}\sum_{t=1}^{T}\sum_{e=1}^{E}\sum_{m=1}^{M}$$

$$A_{\text{Alleviate-Water-Scarcity-Crisis-System}}\,B_{\text{Alleviate-Water-Scarcity-Crisis-System}};$$

$$\sum_{a=1}^{A}\sum_{l=1}^{L}\sum_{l=1}^{L}\sum_{e=1}^{E}\sum_{v=1}^{V}\sum_{i=1}^{I}\sum_{a=1}^{A}\sum_{t=1}^{T}\sum_{e=1}^{E}\sum_{w=1}^{W}\sum_{a=1}^{A}\sum_{t=1}^{T}\sum_{e=1}^{E}\sum_{r=1}^{R}\sum_{s=1}^{S}\sum_{c=1}^{C}\sum_{a=1}^{A}\sum_{r=1}^{R}\sum_{c=1}^{C}\sum_{i=1}^{I}\sum_{t=1}^{T}\sum_{y=1}^{Y}\sum_{c=1}^{C}\sum_{r=1}^{R}\sum_{i=1}^{I}\sum_{s=1}^{S}\sum_{i=1}^{I}\sum_{s=1}^{S}\sum_{s=1}^{S}\sum_{y=1}^{Y}\sum_{s=1}^{S}\sum_{t=1}^{T}\sum_{e=1}^{E}\sum_{m=1}^{M}$$

$$C_{\text{Alleviate-Water-Scarcity-Crisis-System}}\,D_{\text{Alleviate-Water-Scarcity-Crisis-System}};$$

$$\sum_{a=1}^{A}\sum_{l=1}^{L}\sum_{l=1}^{L}\sum_{e=1}^{E}\sum_{v=1}^{V}\sum_{i=1}^{I}\sum_{a=1}^{A}\sum_{t=1}^{T}\sum_{e=1}^{E}\sum_{w=1}^{W}\sum_{a=1}^{A}\sum_{t=1}^{T}\sum_{e=1}^{E}\sum_{r=1}^{R}\sum_{s=1}^{S}\sum_{c=1}^{C}\sum_{a=1}^{A}\sum_{r=1}^{R}\sum_{c=1}^{C}\sum_{i=1}^{I}\sum_{t=1}^{T}\sum_{y=1}^{Y}\sum_{c=1}^{C}\sum_{r=1}^{R}\sum_{i=1}^{I}\sum_{s=1}^{S}\sum_{i=1}^{I}\sum_{s=1}^{S}\sum_{s=1}^{S}\sum_{y=1}^{Y}\sum_{s=1}^{S}\sum_{t=1}^{T}\sum_{e=1}^{E}\sum_{m=1}^{M}$$

$$E_{\textbf{Alleviate-Water-Scarcity-Crisis-System}} F_{\textbf{Alleviate-Water-Scarcity-Crisis-System}};$$

$$\sum_{a=1}^{A}\sum_{l=1}^{L}\sum_{l=1}^{L}\sum_{e=1}^{E}\sum_{v=1}^{V}\sum_{i=1}^{I}\sum_{a=1}^{A}\sum_{t=1}^{T}\sum_{e=1}^{E}\sum_{w=1}^{W}\sum_{a=1}^{A}\sum_{t=1}^{T}\sum_{e=1}^{E}\sum_{r=1}^{R}\sum_{s=1}^{S}\sum_{c=1}^{C}\sum_{a=1}^{A}\sum_{r=1}^{R}\sum_{c=1}^{C}\sum_{i=1}^{I}\sum_{t=1}^{T}\sum_{y=1}^{Y}\sum_{c=1}^{C}\sum_{r=1}^{R}\sum_{i=1}^{I}\sum_{s=1}^{S}\sum_{i=1}^{I}\sum_{s=1}^{S}\sum_{s=1}^{S}\sum_{y=1}^{Y}\sum_{s=1}^{S}\sum_{t=1}^{T}\sum_{e=1}^{E}\sum_{m=1}^{M}$$

$$G_{\textbf{Alleviate-Water-Scarcity-Crisis-System}} H_{\textbf{Alleviate-Water-Scarcity-Crisis-System}};$$

$$\sum_{a=1}^{A}\sum_{l=1}^{L}\sum_{l=1}^{L}\sum_{e=1}^{E}\sum_{v=1}^{V}\sum_{i=1}^{I}\sum_{a=1}^{A}\sum_{t=1}^{T}\sum_{e=1}^{E}\sum_{w=1}^{W}\sum_{a=1}^{A}\sum_{t=1}^{T}\sum_{e=1}^{E}\sum_{r=1}^{R}\sum_{s=1}^{S}\sum_{c=1}^{C}\sum_{a=1}^{A}\sum_{r=1}^{R}\sum_{c=1}^{C}\sum_{i=1}^{I}\sum_{t=1}^{T}\sum_{y=1}^{Y}\sum_{c=1}^{C}\sum_{r=1}^{R}\sum_{i=1}^{I}\sum_{s=1}^{S}\sum_{i=1}^{I}\sum_{s=1}^{S}\sum_{s=1}^{S}\sum_{y=1}^{Y}\sum_{s=1}^{S}\sum_{t=1}^{T}\sum_{e=1}^{E}\sum_{m=1}^{M}$$

$$I_{\textbf{Alleviate-Water-Scarcity-Crisis-System}} J_{\textbf{Alleviate-Water-Scarcity-Crisis-System}};$$

$$\sum_{a=1}^{A}\sum_{l=1}^{L}\sum_{l=1}^{L}\sum_{e=1}^{E}\sum_{v=1}^{V}\sum_{i=1}^{I}\sum_{a=1}^{A}\sum_{t=1}^{T}\sum_{e=1}^{E}\sum_{w=1}^{W}\sum_{a=1}^{A}\sum_{t=1}^{T}\sum_{e=1}^{E}\sum_{r=1}^{R}\sum_{s=1}^{S}\sum_{c=1}^{C}\sum_{a=1}^{A}\sum_{r=1}^{R}\sum_{c=1}^{C}\sum_{i=1}^{I}\sum_{t=1}^{T}\sum_{y=1}^{Y}\sum_{c=1}^{C}\sum_{r=1}^{R}\sum_{i=1}^{I}\sum_{s=1}^{S}\sum_{i=1}^{I}\sum_{s=1}^{S}\sum_{s=1}^{S}\sum_{y=1}^{Y}\sum_{s=1}^{S}\sum_{t=1}^{T}\sum_{e=1}^{E}\sum_{m=1}^{M}$$

$$K_{\textbf{Alleviate-Water-Scarcity-Crisis-System}} L_{\textbf{Alleviate-Water-Scarcity-Crisis-System}};$$

$$\sum_{a=1}^{A}\sum_{l=1}^{L}\sum_{l=1}^{L}\sum_{e=1}^{E}\sum_{v=1}^{V}\sum_{i=1}^{I}\sum_{a=1}^{A}\sum_{t=1}^{T}\sum_{e=1}^{E}\sum_{w=1}^{W}\sum_{a=1}^{A}\sum_{t=1}^{T}\sum_{e=1}^{E}\sum_{r=1}^{R}\sum_{s=1}^{S}\sum_{c=1}^{C}\sum_{a=1}^{A}\sum_{r=1}^{R}\sum_{c=1}^{C}\sum_{i=1}^{I}\sum_{t=1}^{T}\sum_{y=1}^{Y}\sum_{c=1}^{C}\sum_{r=1}^{R}\sum_{i=1}^{I}\sum_{s=1}^{S}\sum_{i=1}^{I}\sum_{s=1}^{S}\sum_{s=1}^{S}\sum_{y=1}^{Y}\sum_{s=1}^{S}\sum_{t=1}^{T}\sum_{e=1}^{E}\sum_{m=1}^{M}$$

$$M_{\textbf{Alleviate-Water-Scarcity-Crisis-System}} N_{\textbf{Alleviate-Water-Scarcity-Crisis-System}};$$

$$\sum_{a=1}^{A}\sum_{l=1}^{L}\sum_{l=1}^{L}\sum_{e=1}^{E}\sum_{v=1}^{V}\sum_{i=1}^{I}\sum_{a=1}^{A}\sum_{t=1}^{T}\sum_{e=1}^{E}\sum_{w=1}^{W}\sum_{a=1}^{A}\sum_{t=1}^{T}\sum_{e=1}^{E}\sum_{r=1}^{R}\sum_{s=1}^{S}\sum_{c=1}^{C}\sum_{a=1}^{A}\sum_{r=1}^{R}\sum_{c=1}^{C}\sum_{i=1}^{I}\sum_{t=1}^{T}\sum_{y=1}^{Y}\sum_{c=1}^{C}\sum_{r=1}^{R}\sum_{i=1}^{I}\sum_{s=1}^{S}\sum_{i=1}^{I}\sum_{s=1}^{S}\sum_{s=1}^{S}\sum_{y=1}^{Y}\sum_{s=1}^{S}\sum_{t=1}^{T}\sum_{e=1}^{E}\sum_{m=1}^{M}$$

$$O_{\textbf{Alleviate-Water-Scarcity-Crisis-System}} P_{\textbf{Alleviate-Water-Scarcity-Crisis-System}} \tag{10.1}$$

$$\textbf{MAX\_A}_{\textbf{a}};\ \textbf{C}_{\textbf{a}};\ \textbf{E}_{\textbf{a}};\ \textbf{G}_{\textbf{a}};\ \textbf{I}_{\textbf{a}};\ \textbf{K}_{\textbf{a}}\ \text{Maximum profit for planting crop}\ \textbf{a}\ \text{in sub-zones} \tag{10.2}$$

$$\textbf{MAX\_B}_{\textbf{a}};\ \textbf{D}_{\textbf{a}};\textbf{F}_{\textbf{a}};\ \textbf{H}_{\textbf{a}};\ \textbf{J}_{\textbf{a}};\ \textbf{L}_{\textbf{a}};\ \textbf{N}_{\textbf{a}};\ \textbf{P}_{\textbf{a}}\ \text{Maximum land area available for planting crop}\ \textbf{a}\ \text{in sub-zones} \tag{10.3}$$

$$\textbf{MIN\_M}_{\textbf{a}}\ \text{Minimum total water consumption for planting crop}\ \textbf{a} \tag{10.4}$$

$$\textbf{MIN\_O}_{\textbf{a}}\ \text{Minimum total kerosene fuel consumption for planting crop}\ \textbf{a} \tag{10.5}$$

$$\textbf{MIN\_s}_{\textbf{a}}\ \text{Minimum total water consumption for planting crop}\ \textbf{a} \tag{10.6}$$

$$\textbf{MAX\_e}_{\textbf{a}}\ \text{Maximum main crop yieldin planting crop}\ \textbf{a} \tag{10.7}$$

$$\textbf{MAX\_i}_{\textbf{a}}\ \text{Maximum main crop price in planting crop}\ \textbf{a} \tag{10.8}$$

$$\textbf{MAX\_m}_{\textbf{a}}\ \text{Maximum Secondary crop yieldin planting crop}\ \textbf{a} \tag{10.9}$$

$$\textbf{MAX\_e}_{\textbf{a}}\ \text{Maximum labor wages cost in planting crop}\ \textbf{a} \tag{10.10}$$

MIN_w_a Minimum draft animals cost in planting crop **a** (10.11)

MIN_a_a Minimum machinery cost in planting crop **a** (10.12)

MIN_e_a Minimum Seeds cost in planting crop **a** (10.13)

MIN_r_a Minimum manure coast in planting crop **a** (10.14)

MIN_s_a Minimum fertilizers coast in planting crop **a** (10.15)

MIN_c_a Minimum insecticides coast in planting crop **a** (10.16)

MIN_r_a Minimum other expenses coast in planting crop **a** (10.17)

MIN_y_a Minimum total kerosene fuel consumption in planting crop **a** (10.18)

MIN_t_a Minimum crop emission in planting crop **a** (10.19)

Variables

$A_{\textbf{Alleviate-Water-Scarcity-Crisis-System}}$ Estimated profits for planting crop **a** in sub-zone (Old land of delta of Egypt) **l** in sub-season **l** in sub-soil type **e** by total main crop yield **v** with main crop price **i** and secondary crop yield **a** with secondary crop price **t** minus of it total crop production cost include labor wages cost **e** draft animals cost **w** machinery cost **a** include it irrigation cost **t** and seeds cost **e** manure coast **r** fertilizers coast **s** insecticides coast **c** laser land leveling **a** other expenses coast **r** and rent cost **c** by total energy consumptions include energy consumption for irrigation **i** energy consumption for labor **t** energy consumption for draft animal **y** energy consumption for land preparation **c** energy consumption for seed planting **r** energy consumption for manure **i** energy consumption for fertilization **s** energy consumption for insecticide **i** energy consumption for laser land leveling **s** and energy consumption for other expenses **s** by kerosene fuel **y** crop water consumption **s** crop emission **t** to give main crop yield **e** and secondary crop yield **m**.

$B_{\textbf{Alleviate-Water-Scarcity-Crisis-System}}$ Estimated land area allocated for planting crop **a** in sub-zone (Old land of delta of Egypt) **l** in sub-season **l** in sub-soil type **e** by total main crop yield **v** with main crop price **i** and secondary crop yield **a** with secondary crop price **t** minus of it total crop production cost include labor wages cost **e** draft animals cost **w** machinery cost **a** include it irrigation cost **t** and seeds cost **e** manure coast **r** fertilizers coast **s** insecticides coast **c** laser land leveling **a** other expenses coast **r** and rent cost **c** by total energy consumptions include energy consumption for irrigation **i** energy consumption for labor **t** energy consumption for draft animal **y** energy consumption for land preparation **c** energy consumption for seed planting **r** energy consumption for manure **i** energy consumption for fertilization **s** energy consumption for insecticide **i** energy consumption for laser land leveling **s** and energy consumption for other expenses **s** by kerosene fuel **y** crop water consumption **s** crop emission **t** to give main crop yield **e** and secondary crop yield **m.**

$C_{\textbf{Alleviate-Water-Scarcity-Crisis-System}}$ Estimated profits for planting crop **a** in sub-zone (Old land of delta of Egypt) **l** in sub-season **l** in sub-soil type **e** by total main crop yield **v** with main crop price **i** and secondary crop yield **a** with secondary crop price **t** minus of it total crop production cost include labor wages cost **e** draft animals cost **w** machinery cost **a** include it irrigation cost **t** and seeds cost **e** manure coast **r** fertilizers coast **s** insecticides coast **c** laser land leveling **a** other expenses coast **r** and rent cost **c** by total energy consumptions include energy consumption for irrigation **i** energy consumption for labor **t** energy consumption for draft animal **y** energy consumption for land preparation **c** energy consumption for seed planting **r** energy consumption for manure **i** energy consumption for fertilization **s** energy consumption for insecticide **i** energy consumption for laser land leveling **s** and energy consumption for other expenses **s** by kerosene fuel **y** crop water consumption **s** crop emission **t** to give main crop yield **e** and secondary crop yield **m.**

$D_{\textbf{Alleviate-Water-Scarcity-Crisis-System}}$ Estimated land area allocated for planting crop **a** in sub-zone (Old land of delta of Egypt) **l** in sub-season **l** in sub-soil type **e** by total main crop yield **v** with main crop price **i** and secondary crop yield **a** with secondary crop price **t** minus of it total crop production cost include labor wages cost **e** draft animals cost **w** machinery cost **a** include it irrigation cost **t** and seeds cost **e** manure coast **r** fertilizers coast **s** insecticides coast **c** laser land leveling **a** other expenses coast **r** and rent cost **c** by total energy consumptions include energy consumption for irrigation **i** energy consumption for labor **t** energy consumption for draft animal **y** energy consumption for land preparation **c** energy consumption for seed planting **r** energy consumption for manure **i** energy consumption for fertilization **s** energy consumption for insecticide **i** energy consumption for laser land leveling **s** and energy consumption for other expenses **s** by kerosene fuel **y** crop water consumption **s** crop emission **t** to give main crop yield **e** and secondary crop yield **m.**

$E_{\textbf{Alleviate-Water-Scarcity-Crisis-System}}$ Estimated profits for planting crop **a** in sub-zone (Old land of delta of Egypt) **l** in sub-season **l** in sub-soil type **e** by total main crop yield **v** with main crop price **i** and secondary crop yield **a** with secondary crop price **t** minus of it total crop production cost include labor wages cost **e** draft animals cost **w** machinery cost **a** include it irrigation cost **t** and seeds cost **e** manure coast **r** fertilizers coast **s** insecticides coast **c** laser land leveling **a** other expenses coast **r** and rent cost **c** by total energy consumptions include energy consumption for irrigation **i** energy consumption for labor **t** energy consumption for draft animal **y** energy consumption for land preparation **c** energy consumption for seed planting **r** energy consumption for manure **i** energy consumption for fertilization **s** energy consumption for insecticide **i** energy consumption for laser land leveling **s** and energy consumption for other expenses **s** by kerosene fuel **y** crop water consumption **s** crop emission **t** to give main crop yield **e** and secondary crop yield **m.**

$F_{\textbf{Alleviate-Water-Scarcity-Crisis-System}}$ Estimated land area allocated for planting crop **a** in sub-zone (Old land of delta of Egypt) **l** in sub-season **l** in sub-soil type **e** by total main crop yield **v** with main crop price **i** and secondary crop yield **a** with secondary crop price **t** minus of it total crop production cost include labor wages cost **e** draft animals cost **w** machinery cost **a** include it irrigation cost **t** and seeds cost **e** manure coast **r** fertilizers coast **s** insecticides coast **c** laser land leveling **a** other expenses coast **r** and rent cost **c** by total energy consumptions include energy consumption for irrigation **i** energy consumption for labor **t** energy consumption for draft animal **y** energy consumption for land preparation **c** energy consumption for seed planting **r** energy consumption for manure **i** energy consumption for fertilization **s** energy consumption for insecticide **i** energy consumption for laser land leveling **s** and energy consumption for other expenses **s** by kerosene fuel **y** crop water consumption **s** crop emission **t** to give main crop yield **e** and secondary crop yield **m**.

$G_{\textbf{Alleviate-Water-Scarcity-Crisis-System}}$ Estimated profits for planting crop **a** in sub-zone (Old land of delta of Egypt) **l** in sub-season **l** in sub-soil type **e** by total main crop yield **v** with main crop price **i** and secondary crop yield **a** with secondary crop price **t** minus of it total crop production cost include labor wages cost **e** draft animals cost **w** machinery cost **a** include it irrigation cost **t** and seeds cost **e** manure coast **r** fertilizers coast **s** insecticides coast **c** laser land leveling **a** other expenses coast **r** and rent cost **c** by total energy consumptions include energy consumption for irrigation **i** energy consumption for labor **t** energy consumption for draft animal **y** energy consumption for land preparation **c** energy consumption for seed planting **r** energy consumption for manure **i** energy consumption for fertilization **s** energy consumption for insecticide **i** energy consumption for laser land leveling **s** and energy consumption for other expenses **s** by kerosene fuel **y** crop water consumption **s** crop emission **t** to give main crop yield **e** and secondary crop yield **m**.

$H_{\textbf{Alleviate-Water-Scarcity-Crisis-System}}$ Estimated land area allocated for planting crop **a** in sub-zone (Old land of delta of Egypt) **l** in sub-season **l** in sub-soil type **e** by total main crop yield **v** with main crop price **i** and secondary crop yield **a** with secondary crop price **t** minus of it total crop production cost include labor wages cost **e** draft animals cost **w** machinery cost **a** include it irrigation cost **t** and seeds cost **e** manure coast **r** fertilizers coast **s** insecticides coast **c** laser land leveling **a** other expenses coast **r** and rent cost **c** by total energy consumptions include energy consumption for irrigation **i** energy consumption for labor **t** energy consumption for draft animal **y** energy consumption for land preparation **c** energy consumption for seed planting **r** energy consumption for manure **i** energy consumption for fertilization **s** energy consumption for insecticide **i** energy consumption for laser land leveling **s** and energy consumption for other expenses **s** by kerosene fuel **y** crop water consumption **s** crop emission **t** to give main crop yield **e** and secondary crop yield **m**.

$I_{\textbf{Alleviate-Water-Scarcity-Crisis-System}}$ Estimated profits for planting crop **a** in sub-zone (Old land of delta of Egypt) **l** in sub-season **l** in sub-soil type **e** by total main crop yield **v** with main crop price **i** and secondary crop yield **a** with secondary crop price **t** minus of it total crop production cost include labor wages cost **e** draft animals cost **w** machinery cost **a** include it irrigation cost **t** and seeds cost **e** manure coast **r** fertilizers coast **s** insecticides coast **c** laser land leveling **a** other expenses coast **r** and rent cost **c** by total energy consumptions include energy consumption for irrigation **i** energy consumption for labor **t** energy consumption for draft animal

y energy consumption for land preparation **c** energy consumption for seed planting **r** energy consumption for manure **i** energy consumption for fertilization **s** energy consumption for insecticide **i** energy consumption for laser land leveling **s** and energy consumption for other expenses **s** by kerosene fuel **y** crop water consumption **s** crop emission **t** to give main crop yield **e** and secondary crop yield **m**.

$J_{\mathbf{Alleviate\text{-}Water\text{-}Scarcity\text{-}Crisis\text{-}System}}$ Estimated land area allocated for planting crop **a** in sub-zone (Old land of delta of Egypt) **l** in sub-season **l** in sub-soil type **e** by total main crop yield **v** with main crop price **i** and secondary crop yield **a** with secondary crop price **t** minus of it total crop production cost include labor wages cost **e** draft animals cost **w** machinery cost **a** include it irrigation cost **t** and seeds cost **e** manure coast **r** fertilizers coast **s** insecticides coast **c** laser land leveling **a** other expenses coast **r** and rent cost **c** by total energy consumptions include energy consumption for irrigation **i** energy consumption for labor **t** energy consumption for draft animal **y** energy consumption for land preparation **c** energy consumption for seed planting **r** energy consumption for manure **i** energy consumption for fertilization **s** energy consumption for insecticide **i** energy consumption for laser land leveling **s** and energy consumption for other expenses **s** by kerosene fuel **y** crop water consumption **s** crop emission **t** to give main crop yield **e** and secondary crop yield **m**.

$K_{\mathbf{Alleviate\text{-}Water\text{-}Scarcity\text{-}Crisis\text{-}System}}$ Estimated profits for planting crop **a** in sub-zone (Old land of delta of Egypt) **l** in sub-season **l** in sub-soil type **e** by total main crop yield **v** with main crop price **i** and secondary crop yield **a** with secondary crop price **t** minus of it total crop production cost include labor wages cost **e** draft animals cost **w** machinery cost **a** include it irrigation cost **t** and seeds cost **e** manure coast **r** fertilizers coast **s** insecticides coast **c** laser land leveling **a** other expenses coast **r** and rent cost **c** by total energy consumptions include energy consumption for irrigation **i** energy consumption for labor **t** energy consumption for draft animal **y** energy consumption for land preparation **c** energy consumption for seed planting **r** energy consumption for manure **i** energy consumption for fertilization **s** energy consumption for insecticide **i** energy consumption for laser land leveling **s** and energy consumption for other expenses **s** by kerosene fuel **y** crop water consumption **s** crop emission **t** to give main crop yield **e** and secondary crop yield **m**.

$L_{\mathbf{Alleviate\text{-}Water\text{-}Scarcity\text{-}Crisis\text{-}System}}$ Estimated land area allocated for planting crop **a** in sub-zone (Old land of delta of Egypt) **l** in sub-season **l** in sub-soil type **e** by total main crop yield **v** with main crop price **i** and secondary crop yield **a** with secondary crop price **t** minus of it total crop production cost include labor wages cost **e** draft animals cost **w** machinery cost **a** include it irrigation cost **t** and seeds cost **e** manure coast **r** fertilizers coast **s** insecticides coast **c** laser land leveling **a** other expenses coast **r** and rent cost **c** by total energy consumptions include energy consumption for irrigation **i** energy consumption for labor **t** energy consumption for draft animal **y** energy consumption for land preparation **c** energy consumption for seed planting **r** energy consumption for manure **i** energy consumption for fertilization **s** energy consumption for insecticide **i** energy consumption for laser land leveling **s** and energy consumption for other expenses **s** by kerosene fuel **y** crop water consumption **s** crop emission **t** to give main crop yield **e** and secondary crop yield **m**.

$M_{\text{Alleviate-Water-Scarcity-Crisis-System}}$ Estimated profits for planting crop **a** in sub-zone (Old land of delta of Egypt) **l** in sub-season **l** in sub-soil type **e** by total main crop yield **v** with main crop price **i** and secondary crop yield **a** with secondary crop price **t** minus of it total crop production cost include labor wages cost **e** draft animals cost **w** machinery cost **a** include it irrigation cost **t** and seeds cost **e** manure coast **r** fertilizers coast **s** insecticides coast **c** laser land leveling **a** other expenses coast **r** and rent cost **c** by total energy consumptions include energy consumption for irrigation **i** energy consumption for labor **t** energy consumption for draft animal **y** energy consumption for land preparation **c** energy consumption for seed planting **r** energy consumption for manure **i** energy consumption for fertilization **s** energy consumption for insecticide **i** energy consumption for laser land leveling **s** and energy consumption for other expenses **s** by kerosene fuel **y** crop water consumption **s** crop emission **t** to give main crop yield **e** and secondary crop yield **m**.

$N_{\text{Alleviate-Water-Scarcity-Crisis-System}}$ Estimated land area allocated for planting crop **a** in sub-zone (Old land of delta of Egypt) **l** in sub-season **l** in sub-soil type **e** by total main crop yield **v** with main crop price **i** and secondary crop yield **a** with secondary crop price **t** minus of it total crop production cost include labor wages cost **e** draft animals cost **w** machinery cost **a** include it irrigation cost **t** and seeds cost **e** manure coast **r** fertilizers coast **s** insecticides coast **c** laser land leveling **a** other expenses coast **r** and rent cost **c** by total energy consumptions include energy consumption for irrigation **i** energy consumption for labor **t** energy consumption for draft animal **y** energy consumption for land preparation **c** energy consumption for seed planting **r** energy consumption for manure **i** energy consumption for fertilization **s** energy consumption for insecticide **i** energy consumption for laser land leveling **s** and energy consumption for other expenses **s** by kerosene fuel **y** crop water consumption **s** crop emission **t** to give main crop yield **e** and secondary crop yield **m**.

$O_{\text{Alleviate-Water-Scarcity-Crisis-System}}$ Estimated profits for planting crop **a** in sub-zone (Old land of delta of Egypt) **l** in sub-season **l** in sub-soil type **e** by total main crop yield **v** with main crop price **i** and secondary crop yield **a** with secondary crop price **t** minus of it total crop production cost include labor wages cost **e** draft animals cost **w** machinery cost **a** include it irrigation cost **t** and seeds cost **e** manure coast **r** fertilizers coast **s** insecticides coast **c** laser land leveling **a** other expenses coast **r** and rent cost **c** by total energy consumptions include energy consumption for irrigation **i** energy consumption for labor **t** energy consumption for draft animal **y** energy consumption for land preparation **c** energy consumption for seed planting **r** energy consumption for manure **i** energy consumption for fertilization **s** energy consumption for insecticide **i** energy consumption for laser land leveling **s** and energy consumption for other expenses **s** by kerosene fuel **y** crop water consumption **s** crop emission **t** to give main crop yield **e** and secondary crop yield **m**.

$P_{\text{Alleviate-Water-Scarcity-Crisis-System}}$ Estimated land area allocated for planting crop **a** in sub-zone (Old land of delta of Egypt) **l** in sub-season **l** in sub-soil type **e** by total main crop yield **v** with main crop price **i** and secondary crop yield **a** with secondary crop price **t** minus of it total crop production cost include labor wages cost **e** draft animals cost **w** machinery cost **a** include it irrigation cost **t** and seeds cost **e** manure coast **r** fertilizers coast **s** insecticides coast **c** laser land leveling **a** other expenses coast **r** and rent cost **c** by total energy consumptions include energy consumption for irrigation **i** energy consumption for labor **t** energy consumption for draft animal **y** energy consumption for land preparation **c** energy consumption for seed planting **r** energy consumption for manure **i** energy consumption for fertilization **s** energy consumption for insecticide **i** energy consumption for laser land leveling **s** and energy consumption for other expenses **s** by kerosene fuel **y** crop water consumption **s** crop emission **t** to give main crop yield **e** and secondary crop yield **m**.

Results of the Model and Discuss It

Alleviate Water Scarcity Crisis System (**AWSCS**) is a model formulated as an analytical tool to apply alleviate water scarcity crisis system in the three zones of old and new land of Egypt in agriculture after laser land leveling did in some fields in the zone and in the season in the sector under the limitations of water sources in Egypt (Fig. 10.2). In addition, the model has the flexibility of introducing alleviate water scarcity crisis system as a pre-requisite to alleviate water scarcity crisis, achieve efficiency and equity in agriculture sector under the global financial and climate change, and as an analytical tool to focus on the scientific linkage between water, food security and employment, reduce cost to become competitive in the world market, reduce water consumption and reduce social cost of pollutants on farming crops. The economic, financial, risk, and the annual internal rate of return analysis of crops production are also investigated. Several steps were followed to implement **AWSCS** model: first step was optimum cropping pattern for each season for every zone in old and new land of Egypt, second step simulated optimum cropping pattern for the three zones (North, Middle and South Egypt), third step simulated optimum cropping pattern for the three zones with existing cropping pattern (2011/2012–2013/2014) to reallocate crop acreage according to efficiency

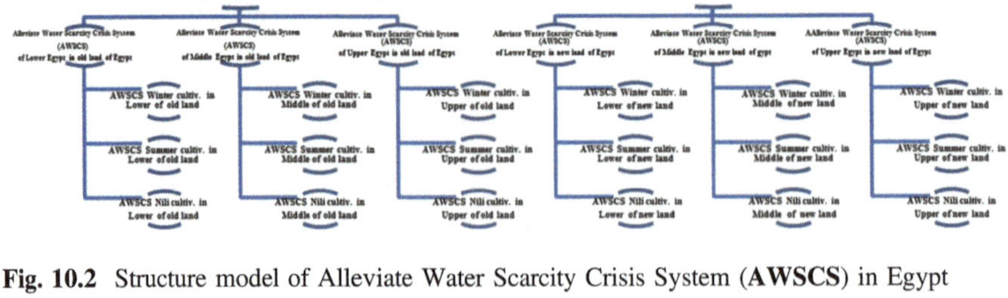

Fig. 10.2 Structure model of Alleviate Water Scarcity Crisis System (**AWSCS**) in Egypt

use of water. To populate the model, field data reported by farmers were used. The data required were collected through a comprehensive survey of water consumption and other inputs to crops fields on a seasonal basis, and included a comprehensive data set relating to the farm enterprise and associated socio-economic conditions. Cropped area, yield, and cost data were obtained from the MALR (2015). Data of water consumption were collected from the MWRI (2015). Necessary data pertaining to cropping pattern input of the respective production system were collected from primary sources and converted into corresponding cropping pattern values. Greenhouse gases emissions were calculated and represented per unit of the energy input. Data presented in this study are representative of typical and/or average data recorded over the three consecutive years of 2011/2012-2013/2014. The existing cultivation and it's an economic evaluation in Egypt in the three regions and seasons in old and new land are presented in Tables 10.1 and 10.2. Remaining base year data are available from the authors and Tables 10.3 and 10.4 places of crops in a larger view, showing place crops and its area as well as cultivating from its source (ECAPMS 2015).

Optimal solutions: Alleviate Water Scarcity Crisis System (AWSCS) is a model should be used. In order to suitable soil type and water could be reallocated to increase farm income, the model adjusted whatever change in land was needed to accompany the changes in soil type and water after made laser land leveling in old and new land of Nile valley. The model structure to optimal cultivation based on suitable soil type and water in Egypt is given in Fig. 10.2. Moreover, data in Tables 10.1 and 10.2 indicate that crop yields of the considered scenario as homogeneous character is higher than their heterogeneous one. In addition, the water requirement rates of the homogeneous consideration were less than the heterogeneous one. Tables 10.3 and 10.4 show economic evaluations for optimum cultivation based on suitable soil type, laser land leveling in old and new land of Nile valley and water and it is comparison with existing condition of Egypt. Figures 10.3 and 10.4 show changes in crops area aggregates in the three zones and seasons from mean 2011/2012–2013/2014 to **AWSCS** in old and new land of Egypt. The results show that area of crops would be 5419.251 and 703.581 hectares cultivated in old and new land of Egypt, respectively, and the proposed model provided higher net benefit than the existing model for all cases. The sum of net benefit for heterogeneous case (342614.709 and 32442.008 Million E.P.) was higher than the sum of homogeneous case (220095.868 and 22072.840 Million E. P.) in old and new land of Egypt and the sum of crop water consumption for heterogeneous case (28538.383 and 3721.615 Million cub. m.) was lower than the sum of homogeneous case (46267.812 and 6351.040 Million cub. m.). It indicates that the variation of heterogeneous character had a large impact on the optimal solution. For this reason, **AWSCS** model with heterogeneous character of land area was appropriate for finding cultivation based on suitable soil type after laser land leveling in old and new land of Nile valley in Egypt has done. An agro-climatic adaptability classification (for each crop) should be established in a form suitable for matching crops with climate and soil resources and crop production cost established according to soil and climatic zone, sufficient to judge whether yields

Table 10.1 Changes area in old land zones flow values from mean (2011/2012–2013/2014) to AWSCS (bold is values that have increased, italic is values that have decreased)

Winter cultivation	Winter cultivation in Lower Egypt				Winter cultivation in Middle Egypt				Winter cultivation in Upper Egypt				Winter cultivation in Egypt			
	Mean	AWSCS	Change	%	Mean	AWSCS	Change	%	Mean	AWSCS	Change	%	Mean	AWSCS	Change	%
Wheat	666456.0	**671994.5**	**5538.5**	**0.8**	218744.8	*211935.4*	*−6809.5*	*−3.1*	182755.9	*179789.0*	*−2966.9*	*−1.6*	**#######**	**#######**	*−4237.80*	*−0.40*
Broad Beans	22649.8	*20153.7*	*−2496.1*	*−11.0*	1865.2	*1594.3*	*−270.9*	*−14.5*	3321.4	**4477.6**	**1156.3**	**34.8**	27836.3	*26225.6*	*−1610.70*	*−5.79*
Barley	4258.4	**4258.4**	**0.0**	**0.0**	1894.6	**2068.5**	**173.9**	**9.2**	197.0	**233.5**	**36.5**	**18.6**	**6350.0**	**6560.4**	**210.42**	**3.31**
Lentil	179.8	**179.8**	**0.0**	**0.0**	0.0	**688.0**	**688.0**	**0.0**	153.7	**688.0**	**534.2**	**347.5**	**333.5**	**1555.7**	**1222.20**	**366.50**
Fenugreek	23.9	**23.9**	**0.0**	**0.0**	1042.9	**1971.1**	**928.2**	**89.0**	648.9	**1133.6**	**484.7**	**74.7**	**1715.7**	**3128.6**	**1412.88**	**82.35**
Chick Peas	0.0	**0.0**	**0.0**	**0.0**	245.3	**388.9**	**143.6**	**0.0**	1189.9	**2776.6**	**1586.8**	**0.0**	**1435.1**	**3165.5**	**1730.40**	**120.57**
Lupine	337.7	**337.7**	**0.0**	**0.0**	5.0	**64.3**	**59.2**	**1175.0**	124.3	**138.2**	**13.9**	**11.1**	**467.0**	**540.1**	**73.08**	**15.65**
Flax	4054.7	**4054.7**	**0.0**	**0.0**	0.0	**0.0**	**0.0**	**0.0**	0.0	**0.0**	**0.0**	**0.0**	**4054.7**	**4054.7**	**0.00**	**0.00**
Onion	24357.5	*24141.6*	*−215.9*	*−0.9*	9533.2	*8095.1*	*−1438.1*	*−15.1*	5966.1	*4234.9*	*−1731.2*	*−29.0*	39856.7	*36471.5*	*−3385.20*	*−8.49*
Clover	353264.9	**353264.9**	**0.0**	**0.0**	115989.3	**129571.7**	**13582.4**	**11.7**	61279.7	**72838.1**	**11558.4**	**18.9**	530533.9	**555674.7**	**25140.78**	**4.74**
Clover Tahreesh	98081.8	**98081.8**	**0.0**	**0.0**	20372.1	*19936.1*	*−436.0*	*−2.1*	5836.3	*5741.8*	*−94.5*	*−1.6*	124290.2	*123759.7*	*−530.46*	*−0.43*
Garlic	1351.1	**1351.1**	**0.0**	**0.0**	8181.2	*7990.5*	*−190.7*	*−2.3*	807.2	**851.8**	**44.5**	**5.5**	10339.6	*10193.4*	*−146.16*	*−1.41*
Sugar Beet	90047.2	**90047.2**	**0.0**	**0.0**	29574.3	**30867.9**	**1293.6**	**4.4**	2504.0	*2048.8*	*−455.3*	*−18.2*	122125.5	**122963.8**	**838.32**	**0.69**
Tomato	17569.0	**17569.0**	**0.0**	**0.0**	14487.1	**15586.6**	**1099.6**	**7.6**	9916.6	*9453.8*	*−462.8*	*−4.7*	41972.7	**42609.4**	**636.72**	**1.52**
Vegetables	116413.1	*75637.4*	*−40775.7*	*−35.0*	32676.0	*27087.5*	*−5588.5*	*−17.1*	4009.7	**4105.1**	**95.3**	**2.4**	153098.8	*106829.9*	*−46268.9*	*−30.22*
Others	1717.0	**797815.6**	**796098.7**	**46366.8**	13635.7	**15688.3**	**2052.5**	**15.1**	1247.8	**1531.7**	**283.9**	**22.8**	16600.5	**815035.6**	**798435.1**	**4809.71**
Summer cultivation	Summer cultivation in Lower Egypt				Summer cultivation in Middle Egypt				Summer cultivation in Upper Egypt				Summer cultivation in Egypt			
	Mean	AWSCS	Change	%	Mean	AWSCS	Change	%	Mean	AWSCS	Change	%	Mean	AWSCS	Change	%
Cotton	99240.1	**192531.4**	**93291.2**	**94.0**	15632.4	*11771.8*	*−3860.6*	*−24.7*	4776.7	*1445.2*	*−3331.4*	*−69.7*	119649.2	**205748.3**	**86099.16**	**71.96**
Rice	558272.8	**549007.6**	**9245.5**	**1.7**	119.7	**637.1**	**481.3**	**402.1**	0.0	**0.0**	**0.0**	**0.0**	558392.5	*549644.8*	*−8747.76*	*−1.57*
Sorghum	0.0	**0.0**	**0.0**	**0.0**	45231.1	**48860.3**	**5050.9**	**11.2**	101077.2	**79094.0**	**8016.5**	**9.6**	146308.3	*127954.3*	*#######*	*−12.54*
Maize	310878.1	**296384.8**	**9834.7**	**3.3**	184075.5	**220580.9**	**57306.1**	**0.0**	94768.0	*119527.8*	*−15493.0*	*−13.7*	589721.6	**636493.5**	**46771.90**	**7.93**
Corn	71796.1	*42100.8*	*−4075.3*	*−9.7*	15164.1	*20107.8*	*−1553.6*	*−10.2*	32386.2	**44095.4**	**1381.8**	**3.6**	119346.4	*106304.0*	*#######*	*−10.93*
Soya Bean	113.4	*316.7*	*−152.5*	*0.0*	8556.2	*7883.8*	*−2093.3*	*−24.5*	607.3	**520.8**	**45.4**	**7.9**	9277.0	*8721.3*	*−555.66*	*−5.99*
Sugar Cane	1046.2	*1476.3*	*−188.6*	*−12.8*	17719.0	*11525.6*	*−118.4*	*−0.7*	100868.5	*101831.9*	*−194.5*	*−0.2*	119633.6	*114833.9*	*−4799.76*	*−4.01*
Peanut	10928.0	*12878.5*	*−4229.8*	*−32.8*	5124.4	*4376.0*	*−968.5*	*0.0*	1727.5	**982.0**	**252.8**	**0.0**	17779.9	**18236.4**	**456.54**	**2.57**
Potatoes	38255.3	*47717.5*	*−10971.2*	*−23.0*	5625.9	**6470.7**	**1718.2**	**30.5**	749.7	**173.9**	**198.7**	**115.1**	44630.9	**54362.0**	**9731.12**	**21.80**

(continued)

Table 10.1 (continued)

Winter cultivation	Winter cultivation in Lower Egypt				Winter cultivation in Middle Egypt				Winter cultivation in Upper Egypt				Winter cultivation in Egypt			
	Mean	AWSCS	Change	%	Mean	AWSCS	Change	%	Mean	AWSCS	Change	%	Mean	AWSCS	Change	%
Onion	735.4	1593.5	366.7	23.0	2686.7	1653.4	−1648.5	−67.6	0.0	0.0	0.0	0.0	3422.2	3246.9	−175.28	−5.12
Sesame	3976.1	8012.8	−2999.2	−37.4	6423.5	5332.5	−1862.3	−29.0	2873.6	1359.5	1755.6	84.5	13273.3	14704.8	1431.50	10.78
Sunflower	1701.8	2011.8	236.9	11.8	3027.4	2276.1	−794.2	−26.2	1033.6	367.5	1098.7	84.7	5762.8	4655.4	−1107.40	−19.22
Tomato	26646.5	23124.8	5036.6	21.8	12975.5	12368.3	978.2	7.5	3510.4	2990.8	−950.0	−25.2	43132.3	38483.9	−4648.42	−10.78
Vegetables	151271.0	117352.6	−178.9	−0.2	32109.0	27837.3	8826.3	27.5	7726.7	8126.2	3508.7	59.2	191106.7	153316.1	#######	−19.77
Others	128733.4	162360.7	111952.3	68.9	45746.4	42075.3	17284.7	61.7	12865.4	12081.7	2166.4	16.3	187345.2	216517.7	29172.50	15.57
Nili cultivation	Nili cultivation in Lower Egypt				Nili cultivation in Middle Egypt				Nili cultivation in Upper Egypt				Nili cultivation in Egypt			
	Mean	AWSCS	Change	%	Mean	AWSCS	Change	%	Mean	AWSCS	Change	%	Mean	AWSCS	Change	%
Rice	0.00	0.00	0.00	0.00	70.6	29.40	−164.22	−100.00	0.00	0.00	0.00	0.00	70.6	29.4	−41.16	−58.33
Sorghum	0.00	0.00	0.00	0.00	802.5	1425.90	−1036.98	−100.00	18.90	8.82	−8.82	−100.00	821.5	1434.7	613.20	74.64
Maize	30690.66	42814.38	6878.34	19.14	45719.1	58225.02	−2782.08	−6.10	8687.70	17455.62	25358.76	145.28	85097.5	118495.0	33397.56	39.25
Corn	22646.40	19212.90	1015.98	5.58	0.0	0.00	19074.72	13804.26	3494.40	1647.66	17565.24	1066.07	26140.8	20860.6	−5280.24	−20.20
Onion	0.00	0.00	0.00	0.00	3120.2	4019.82	−2405.34	−100.00	733.32	808.92	−808.92	−100.00	3853.5	4828.7	975.24	25.31
Potatoes	4730.46	4709.04	186.06	4.11	15716.4	16604.70	#######	−72.93	776.16	702.66	4006.38	570.17	21223.0	22016.4	793.38	3.74
Tomato	1785.00	2134.86	271.32	14.56	10699.5	9896.04	−8449.98	−79.83	700.56	1420.02	714.84	50.34	13185.1	13450.9	265.86	2.02
Vegetables	17823.54	19054.56	6347.46	49.95	11686.9	11008.62	6196.68	48.19	754.74	1538.46	17516.10	1138.55	30265.2	31601.6	1336.44	4.42
Others	6433.56	16462.32	1807.68	12.34	6046.7	12608.82	6791.82	70.23	1343.16	1751.40	14710.92	839.95	13823.5	30822.5	16999.08	122.97

Data source (1) MALR (2015) (2) AWSCS model (2015)

Table 10.2 Changes area in new land zones flow values from mean (2011/2012–2013/2014) to AWSCS (bold is values that have increased, italic is values that have decreased)

Winter cultivation	Winter cultivation in Lower Egypt				Winter cultivation in Middle Egypt				Winter cultivation in Upper Egypt				Winter cultivation in Egypt			
	Mean	AWSCS	Change	%	Mean	AWSCS	Change	%	Mean	AWSCS	Change	%	Mean	AWSCS	Change	%
Wheat	68931.7	**84000.0**	**15068.3**	**21.9**	20480.5	*18557.3*	*−1923.2*	*−9.4*	42232.7	**54406.8**	**12174.1**	**28.8**	131644.8	**156964.1**	**25319.28**	**19.23**
Broad Beans	4197.1	**6062.7**	**1865.6**	**44.5**	25.6	**281.4**	**255.8**	**998.4**	1097.0	*821.5*	*−275.5*	*−25.1*	5319.7	**7165.6**	**1845.90**	**34.70**
Barley	7095.1	**9802.4**	**2707.3**	**38.2**	371.7	**963.5**	**591.8**	**159.2**	4621.7	*881.2*	*−3740.5*	*−80.9*	12088.4	*11647.0*	*−441.42*	*−3.65*
Lentil	0.0	**0.0**	**0.0**	**0.0**	0.0	**6.3**	**6.3**	**0.0**	6.3	*0.0*	*−6.3*	*−100.0*	6.3	**6.3**	**0.00**	**0.00**
Fenugreek	24.4	**27.3**	**2.9**	**12.1**	113.0	**364.1**	**251.2**	**222.3**	59.6	**142.8**	**83.2**	**139.4**	197.0	**534.2**	**337.26**	**171.22**
Chick Peas	0.0	**0.0**	**0.0**	**0.0**	0.0	**0.0**	**0.0**	**0.0**	0.4	*0.0*	*−0.4*	*0.0*	0.4	*0.0*	*−0.42*	*−100.00*
Lupine	276.4	*262.9*	*−13.4*	*−4.9*	0.4	*0.0*	*−0.4*	*−100.0*	26.0	*10.5*	*−15.5*	*−59.7*	302.8	*273.4*	*−29.40*	*−9.71*
Flax	31.1	**105.8**	**74.8**	**240.5**	0.0	**0.0**	**0.0**	**0.0**	0.0	**0.0**	**0.0**	**0.0**	31.1	**105.8**	**74.76**	**240.54**
Onion	1977.8	*1885.8*	*−92.0*	*−4.7*	3363.4	**4286.9**	**923.6**	**27.5**	5161.8	**5936.3**	**774.5**	**15.0**	10502.9	**12109.0**	**1606.08**	**15.29**
Clover	25668.7	**35782.7**	**10114.0**	**39.4**	2853.1	**6085.4**	**3232.3**	**113.3**	5397.8	**6792.7**	**1394.8**	**25.8**	33919.6	**48660.8**	**14741.16**	**43.46**
Clover Tahreesh	10106.9	**10594.9**	**488.0**	**4.8**	105.0	*0.0*	*−105.0*	*−100.0*	0.0	**0.0**	**0.0**	**0.0**	10211.9	**10594.9**	**383.04**	**3.75**
Garlic	301.6	*220.1*	*−81.5*	*−27.0*	42.0	**113.8**	**71.8**	**171.0**	148.3	**226.4**	**78.1**	**52.7**	491.8	**560.3**	**68.46**	**13.92**
Sugar Beet	32658.4	**45939.6**	**13281.2**	**40.7**	2049.6	*1217.6*	*−832.0*	*−40.6*	96.6	**363.3**	**266.7**	**276.1**	34804.6	**47520.5**	**12715.92**	**36.54**
Tomato	10579.4	*10077.9*	*−501.5*	*−4.7*	7540.3	**10166.1**	**2625.8**	**34.8**	10235.8	**14159.0**	**3923.2**	**38.3**	28355.5	**34403.0**	**6047.58**	**21.33**
Vegetables	20891.2	*30476.5*	*9585.2*	*45.9*	3690.5	*3137.8*	*−552.7*	*−15.0*	3993.4	*3155.9*	*−837.5*	*−21.0*	28575.1	**36770.2**	**8195.0**	**28.68**
Others	3058.0	**2448.6**	**−609.4**	**−19.9**	1934.9	**2632.6**	**697.6**	**36.1**	483.8	**707.3**	**223.4**	**46.2**	5476.8	**5788.4**	**311.6**	**5.69**
Summer cultivation	Summer cultivation in Lower Egypt				Summer cultivation in Middle Egypt				Summer cultivation in Upper Egypt				Summer cultivation in Egypt			
	Mean	AWSCS	Change	%	Mean	AWSCS	Change	%	Mean	AWSCS	Change	%	Mean	AWSCS	Change	%
Cotton	9256.4	*1641.8*	*−7614.6*	*−82.3*	0.0	**0.0**	**0.0**	**0.0**	0.0	**0.0**	**0.0**	**0.0**	9256.4	*1641.8*	*−7614.60*	*−82.26*
Rice	35411.5	**41065.9**	**9245.5**	**1.7**	0.0	**0.0**	**481.3**	**402.1**	0.0	**0.0**	**0.0**	**0.0**	35411.5	**41065.9**	**5654.46**	**15.97**
Sorghum	0.0	**0.0**	**0.0**	**0.0**	703.9	**777.0**	**5050.9**	**11.2**	7863.7	**9372.7**	**8016.5**	**9.6**	8567.6	**10149.7**	**1582.14**	**18.47**
Maize	22347.8	**21755.2**	**9834.7**	**3.3**	8649.1	**10954.9**	**57306.1**	**0.0**	1803.9	*4834.6*	*−15493.0*	*−13.7*	32800.7	**37544.6**	**4743.90**	**14.46**
Corn	6848.9	*10243.0*	*−4075.3*	*−9.7*	777.8	*420.8*	*−1553.6*	*−10.2*	3307.1	**6192.1**	**1381.8**	**3.6**	10933.9	**16855.9**	**5922.00**	**54.16**
Soya Bean	8.4	*0.0*	*−152.5*	*0.0*	47.5	*7.6*	*−2093.3*	*−24.5*	0.0	**13.4**	**45.4**	**7.9**	55.9	*21.0*	*−34.86*	*−62.41*
Sugar Cane	10.9	*0.0*	*−188.6*	*−12.8*	45.8	*0.0*	*−118.4*	*−0.7*	16340.1	*18136.4*	*−194.5*	*−0.2*	16396.8	**18136.4**	**1739.64**	**10.61**
Peanut	7968.2	*11678.1*	*−4229.8*	*−32.8*	2872.0	*3869.0*	*−968.5*	*0.0*	1839.2	**1748.0**	**252.8**	**0.0**	12679.4	**17295.2**	**4615.80**	**36.40**
Potatoes	3041.6	*4279.4*	*−10971.2*	*−23.0*	189.0	**186.9**	**1718.2**	**30.5**	0.0	**3444.0**	**198.7**	**115.1**	3230.6	**7910.3**	**4679.64**	**144.85**
Onion	115.5	**10.5**	**366.7**	**23.0**	614.0	*447.3*	*−1648.5*	*−67.6*	0.0	**0.0**	**0.0**	**0.0**	729.5	*457.8*	*−271.74*	*−37.25*
Sesame	2897.2	*5509.1*	*−2999.2*	*−37.4*	1866.5	*2644.7*	*−1862.3*	*−29.0*	2062.2	**3290.3**	**1755.6**	**84.5**	6825.8	**11444.2**	**4618.32**	**67.66**

(continued)

Table 10.2 (continued)

Winter cultivation	Winter cultivation in Lower Egypt				Winter cultivation in Middle Egypt				Winter cultivation in Upper Egypt				Winter cultivation in Egypt			
	Mean	AWSCS	Change	%	Mean	AWSCS	Change	%	Mean	AWSCS	Change	%	Mean	AWSCS	Change	%
Sunflower	120.1	**204.5**	**236.9**	**11.8**	209.6	*260.4*	*−794.2*	*−26.2*	0.0	**1.3**	**1098.7**	**84.7**	329.7	**466.2**	**136.50**	**41.40**
Tomato	18412.4	**23539.3**	**5036.6**	**21.8**	3065.6	**2682.5**	**978.2**	**7.5**	551.5	*1752.7*	*−950.0*	*−25.2*	22029.4	**27974.5**	**5945.10**	**26.99**
Vegetables	35025.9	*62701.4*	*−178.9*	*−0.2*	4957.3	**6444.9**	**8826.3**	**27.5**	3785.5	**5060.2**	**3508.7**	**59.2**	43768.6	**74206.4**	**30437.82**	**69.54**
Others	17227.1	**25309.2**	**111952.3**	**68.9**	930.3	**2793.8**	**17284.7**	**61.7**	7382.3	**10526.9**	**2166.4**	**16.3**	25539.8	**38629.9**	**13090.14**	**51.25**
Nili cultivation	Nili cultivation in Lower Egypt				Nili cultivation in Middle Egypt				Nili cultivation in Upper Egypt				Nili cultivation in Egypt			
	Mean	AWSCS	Change	%	Mean	AWSCS	Change	%	Mean	AWSCS	Change	%	Mean	AWSCS	Change	%
Rice	0.00	**0.00**	**0.00**	**0.00**	0.0	*0.00*	*−164.22*	*−100.00*	0.00	**0.00**	**0.00**	**0.00**	0.0	*0.0*	*0.00*	*0.00*
Sorghum	0.00	**0.00**	**0.00**	**0.00**	0.0	*0.00*	*−1036.98*	*−100.00*	12.60	*0.00*	*−8.82*	*−100.00*	12.6	*0.0*	*−12.60*	*−100.00*
Maize	835.80	**2332.68**	**6878.34**	**19.14**	590.5	*609.00*	*−2782.08*	*−6.10*	1153.74	**2062.62**	**25358.76**	**145.28**	2580.1	**5004.3**	**2424.24**	**93.96**
Corn	2232.30	**3558.24**	**1015.98**	**5.58**	0.0	**0.00**	**19074.72**	**13804.26**	281.40	**2241.96**	**17565.24**	**1066.07**	2513.7	**5800.2**	**3286.50**	**130.74**
Onion	21.00	**75.60**	**0.00**	**0.00**	383.0	*216.30*	*−2405.34*	*−100.00*	513.66	*844.20*	*−808.92*	*−100.00*	917.7	**1136.1**	**218.40**	**23.80**
Potatoes	397.32	**1300.74**	**186.06**	**4.11**	184.8	*184.80*	*−12686.10*	*−72.93*	0.00	**0.00**	**4006.38**	**570.17**	582.1	**1485.5**	**903.42**	**155.19**
Tomato	756.42	**2424.66**	**271.32**	**14.56**	972.7	*1250.76*	*−8449.98*	*−79.83*	656.88	**650.58**	**714.84**	**50.34**	2386.0	**4326.0**	**1939.98**	**81.31**
Vegetables	1887.06	**4362.12**	**6347.46**	**49.95**	739.2	**1543.50**	**6196.68**	**48.19**	847.14	**550.62**	**17516.10**	**1138.55**	3473.4	**6456.2**	**2982.84**	**85.88**
Others	20.16	**1877.40**	**1807.68**	**12.34**	0.0	**361.20**	**6791.82**	**70.23**	395.64	**230.16**	**14710.92**	**839.95**	415.8	**2468.8**	**2052.96**	**493.74**

Data source (1) MALR (2015) (2) AWSCS model (2015)

Table 10.3 Changes in old land area, unit values and aggregate zones flow values from mean (2011/2012–2013/2014) to AWSCS (bold are values that have increased, italic are values that have decreased)

Lower Egypt	Winter cultivation				Summer cultivation				Nili cultivation				Cultivation in Lower Egypt			
	Mean	AWSCS	Change	%	Mean	AWSCS	Change	%	Mean	AWSCS	Change	%	Mean	AWSCS	Change	%
Irrigated area of crop in old land	2922.49	**2922.49**	**0.00**	**0.00**	2253.22	**2253.22**	**0.00**	**0.00**	243.54	**243.54**	**0.00**	**0.00**	5419.25	**5419.25**	**0.00**	**0.00**
Soil type	11.93	*14.61*	*2.67*	*22.38*	9.06	*8.61*	*−0.45*	*−4.97*	1.11	**1.17**	**0.06**	**5.24**	22.10	**24.38**	**2.28**	**10.31**
Main crop yield	103.12	**136.90**	**33.78**	**32.76**	78.22	**96.06**	**17.85**	**22.82**	6.94	**10.49**	**3.55**	**51.14**	188.28	**243.45**	**55.18**	**29.31**
Secondary crop yield	30.41	**34.30**	**3.89**	**12.79**	35.58	**43.73**	**8.15**	**22.90**	2.59	**3.47**	**0.88**	**33.84**	68.58	**81.50**	**12.92**	**18.83**
Main crop price	19253.37	**20964.03**	**1710.66**	**8.88**	6603.15	**8465.56**	**1862.41**	**28.20**	726.43	**766.33**	**39.90**	**5.49**	26582.95	**30195.92**	**3612.97**	**13.59**
Secondary crop price	391.47	**426.56**	**35.10**	**8.97**	119.11	**128.59**	**9.48**	**7.96**	7.13	**10.40**	**3.27**	**45.91**	517.70	**565.55**	**47.85**	**9.24**
Total crop production cost	23410.93	*26607.35*	*3196.42*	*13.65*	24969.43	*26684.42*	*1714.99*	*6.87*	2381.39	*2484.74*	*103.35*	*4.34*	50761.75	*55776.52*	*5014.76*	*9.88*
*Labor Wages	3476.34	**7151.47**	**3675.13**	**105.72**	5525.14	**7397.02**	**1871.88**	**33.88**	491.20	**715.56**	**224.36**	**45.68**	9492.68	**15264.05**	**5771.38**	**60.80**
Draft Animals	30.77	*31.77*	*1.00*	*3.24*	179.90	**79.08**	**−100.81**	**−56.04**	4.19	**2.71**	**−1.49**	**−35.45**	214.86	**113.55**	**−101.31**	**−47.15**
Machinery	2189.90	*2775.31*	*585.41*	*26.73*	3046.96	*3307.32*	*260.36*	*8.54*	238.21	*249.01*	*10.79*	*4.53*	5475.07	*6331.63*	*856.56*	*15.64*
Seeds Cost	1035.28	*1442.39*	*407.11*	*39.32*	1689.77	*1802.09*	*112.32*	*6.65*	200.69	*213.32*	*12.63*	*6.29*	2925.74	*3457.79*	*532.05*	*18.19*
Manure	350.22	*478.36*	*128.15*	*36.59*	762.16	**702.74**	**−59.42**	**−7.80**	100.16	*110.63*	*10.47*	*10.45*	1212.54	*1291.73*	*79.19*	*6.53*
Fertilizers	1869.94	*2644.19*	*774.25*	*41.40*	2678.07	*2738.88*	*60.81*	*2.27*	308.06	**302.20**	**−5.85**	**−1.90**	4856.08	*5685.28*	*829.20*	*17.08*
Insecticides	369.80	*492.54*	*122.74*	*33.19*	439.82	*504.20*	*64.38*	*14.64*	67.20	**59.40**	**−7.80**	**−11.60**	876.82	*1056.14*	*179.32*	*20.45*
Laser land leveling cost	153.77	**153.77**	**0.00**	**0.00**	112.62	**109.56**	**−3.06**	**−2.72**	8.53	*10.49*	*1.96*	*23.00*	274.92	**273.82**	**−1.10**	**−0.40**
Other Expenses	874.16	*1225.88*	*351.73*	*40.24*	1498.35	*1608.22*	*109.86*	*7.33*	139.12	*150.99*	*11.86*	*8.53*	2511.63	*2985.09*	*473.45*	*18.85*
Rent	7206.27	*9629.35*	*2423.1*	*33.62*	8071.75	*8435.33*	*363.58*	*4.50*	652.71	*670.44*	*17.73*	*2.72*	15930.73	*18735.12*	*2804.4*	*17.60*
Crop revenue	157470.67	**257548.65**	**#######**	**63.55**	57960.85	**76210.05**	**18249.20**	**31.49**	4664.35	**8856.01**	**4191.66**	**89.87**	#######	**#######**	**122518.8**	**55.67**
Crop profit	139919.85	**226132.71**	**86212.86**	**61.62**	36205.85	**49525.62**	**13319.77**	**36.79**	3142.74	**4370.17**	**1227.43**	**39.06**	#######	**#######**	**100760.1**	**56.21**

(continued)

Table 10.3 (continued)

Lower Egypt	Winter cultivation				Summer cultivation				Nili cultivation				Cultivation in Lower Egypt			
	Mean	AWSCS	Change	%	Mean	AWSCS	Change	%	Mean	AWSCS	Change	%	Mean	AWSCS	Change	%
Crop water consumption	16487.21	**9828.19**	**−6659.01**	**−40.39**	27955.68	**17438.51**	**#######**	**−37.62**	1824.92	**1271.68**	**−553.24**	**−30.32**	46267.81	**28538.38**	**−17729.4**	**−38.32**
Kerosene fuel Liter	4421.94	**3683.80**	**−738.14**	**−16.69**	2003.59	**1739.89**	**−263.70**	**−13.16**	198.78	**173.87**	**−24.91**	**−12.53**	6624.31	**5597.56**	**−1026.8**	**−15.50**
Energy consumption in cultivation MJ	96355.39	**81642.18**	**#######**	**−15.27**	#######	**94563.56**	**#######**	**−13.84**	10847.58	**9424.51**	**−1423.08**	**−13.12**	#######	**#######**	**−31324.0**	**−14.44**
Middle Egypt	Winter cultivation				Summer cultivation				Nili cultivation				Cultivation in Middle Egypt			
	Mean	AWSCS	Change	%	Mean	AWSCS	Change	%	Mean	AWSCS	Change	%	Mean	AWSCS	Change	%
Irrigated area of crop in old land	473.53	**473.53**	**0.00**	**0.00**	423.76	**423.76**	**0.00**	**0.00**	113.82	**113.82**	**0.00**	**0.00**	1011.11	**1011.11**	**0.00**	**0.00**
Soil type	2.12	**2.12**	**0.00**	**0.20**	2.71	*2.46*	*−0.25*	*−9.11*	0.51	*0.65*	*0.14*	*27.26*	5.34	**5.24**	**−0.10**	**−1.95**
Main crop yield	21.61	**25.89**	**4.28**	**19.80**	16.25	**21.77**	**5.53**	**34.00**	3.24	**4.69**	**1.45**	**44.82**	41.10	**52.36**	**11.26**	**27.39**
Secondary crop yield	6.97	**7.58**	**0.61**	**8.82**	6.78	**9.31**	**2.53**	**37.39**	1.13	**1.47**	**0.34**	**30.25**	14.87	**18.36**	**3.49**	**23.47**
Main crop price	1646.95	**2004.20**	**357.25**	**21.69**	874.55	**1178.96**	**304.41**	**34.81**	327.01	**348.28**	**21.27**	**6.50**	2848.51	**3531.44**	**682.93**	**23.98**
Secondary crop price	67.35	**83.96**	**16.60**	**24.65**	20.94	**25.64**	**4.70**	**22.44**	3.11	**4.41**	**1.30**	**41.82**	91.40	**114.00**	**22.60**	**24.73**
Total crop production cost	3866.07	*4731.81*	*865.74*	*22.39*	4536.64	*4584.14*	*47.51*	*1.05*	1163.69	*1234.15*	*70.46*	*6.05*	9566.39	*10550.09*	*983.70*	*10.28*
*Labor Wages	775.31	**1057.49**	**282.17**	**36.40**	954.12	**1320.89**	**366.76**	**38.44**	251.09	**354.78**	**103.69**	**41.30**	1980.52	**2733.15**	**752.63**	**38.00**
Draft Animals	6.24	*7.14*	*0.90*	*14.47*	24.06	**11.81**	**−12.24**	**−50.89**	1.93	**1.34**	**−0.58**	**−30.37**	32.22	**20.30**	**−11.93**	**−37.01**
Machinery	473.98	*528.18*	*54.20*	*11.44*	514.20	*521.66*	*7.46*	*1.45*	112.25	*118.22*	*5.97*	*5.32*	1100.43	*1168.06*	*67.63*	*6.15*
Seeds Cost	236.30	*262.27*	*25.97*	*10.99*	315.89	**308.61**	**−7.28**	**−2.31**	115.23	*125.31*	*10.08*	*8.75*	667.42	*696.19*	*28.77*	*4.31*
Manure	78.28	*94.91*	*16.64*	*21.25*	182.52	**168.04**	**−14.47**	**−7.93**	44.07	*48.83*	*4.77*	*10.81*	304.86	*311.79*	*6.93*	*2.27*
Fertilizers	420.64	*432.51*	*11.87*	*2.82*	521.92	**495.43**	**−26.50**	**−5.08**	158.31	**156.93**	**−1.38**	**−0.87**	1100.88	**1084.87**	**−16.01**	**−1.45**
Insecticides	81.31	*88.51*	*7.19*	*8.85*	60.88	*78.50*	*17.62*	*28.94*	33.52	**30.42**	**−3.10**	**−9.25**	175.72	*197.43*	*21.71*	*12.36*
Laser land leveling cost	28.63	**28.63**	**0.00**	**0.00**	22.13	**19.34**	**−2.79**	**−12.59**	4.27	*5.12*	*0.86*	*20.05*	55.03	**53.10**	**−1.93**	**−3.51**
Other Expenses	193.59	*218.85*	*25.25*	*13.04*	271.69	*275.54*	*3.86*	*1.42*	68.36	*76.45*	*8.09*	*11.83*	533.64	*570.84*	*37.20*	*6.97*

(continued)

Table 10.3 (continued)

Middle Egypt	Winter cultivation				Summer cultivation				Nili cultivation				Cultivation in Middle Egypt			
	Mean	AWSCS	Change	%	Mean	AWSCS	Change	%	Mean	AWSCS	Change	%	Mean	AWSCS	Change	%
Rent	1557.19	*1715.70*	*158.5*	*10.18*	1410.91	**1384.31**	**−26.60**	**−1.89**	305.92	*316.74*	*10.82*	*3.54*	3274.02	*3416.75*	*142.7*	*4.36*
Crop revenue	30891.89	**38839.05**	**7947.16**	**25.73**	10071.19	**13621.41**	**3550.22**	**35.25**	2545.26	**4337.22**	**1791.96**	**70.40**	43508.34	**56797.68**	**13289.3**	**30.54**
Crop profit	27025.05	**33748.77**	**6723.72**	**24.88**	6582.90	**9037.27**	**2454.36**	**37.28**	1903.64	**2346.35**	**442.71**	**23.26**	35511.60	**45132.39**	**9620.8**	**27.09**
Crop water consumption	2877.61	**1757.29**	**−1120.32**	**−38.93**	4390.49	**2728.08**	**−1662.41**	**−37.86**	837.91	**620.48**	**−217.43**	**−25.95**	8106.01	**5105.85**	**−3000.2**	**−37.01**
Kerosene fuel Liter	651.80	**562.60**	**−89.20**	**−13.69**	380.39	**304.79**	**−75.60**	**−19.87**	95.15	**84.38**	**−10.77**	**−11.32**	1127.34	**951.76**	**−175.6**	**−15.57**
Energy consumption in cultivation MJ	15848.04	**13759.43**	**−2088.60**	**−13.18**	20193.10	**16816.86**	**−3376.24**	**−16.72**	5193.57	**4574.87**	**−618.70**	**−11.91**	41234.70	**35151.16**	**−6083.5**	**−14.75**
Upper Egypt	Winter cultivation				Summer cultivation				Nili cultivation				Cultivation in Upper Egypt			
	Mean	AWSCS	Change	%	Mean	AWSCS	Change	%	Mean	AWSCS	Change	%	Mean	AWSCS	Change	%
Irrigated area of crop in old land	290.04	**290.04**	**0.00**	**0.00**	372.60	**372.60**	**0.00**	**0.00**	25.33	**25.33**	**0.00**	**0.00**	687.97	**687.97**	**0.00**	**0.00**
Soil type	1.14	*1.57*	*0.44*	*38.35*	1.46	**1.44**	**−0.02**	**−1.48**	0.15	**0.11**	**−0.04**	**−27.30**	2.75	*3.12*	*0.37*	*13.56*
Main crop yield	14.46	**17.82**	**3.36**	**23.22**	23.48	**28.94**	**5.46**	**23.24**	0.61	**1.22**	**0.61**	**100.88**	38.55	**47.98**	**9.43**	**24.46**
Secondary crop yield	5.02	**5.65**	**0.63**	**12.55**	5.79	**7.16**	**1.37**	**23.72**	0.25	**0.47**	**0.22**	**86.99**	11.06	**13.29**	**2.22**	**20.08**
Main crop price	735.39	**956.43**	**221.04**	**30.06**	439.55	**507.86**	**68.31**	**15.54**	50.50	**51.48**	**0.99**	**1.95**	1225.44	**1515.78**	**290.34**	**23.69**
Secondary crop price	56.62	**69.27**	**12.65**	**22.34**	17.55	**19.61**	**2.06**	**11.76**	0.69	**1.41**	**0.72**	**104.35**	74.85	**90.29**	**15.43**	**20.62**
Total crop production cost	2477.07	*2992.23*	*515.15*	*20.80*	4388.47	*4563.56*	*175.09*	*3.99*	227.15	*238.13*	*10.98*	*4.83*	7092.70	*7793.92*	*701.22*	*9.89*
*Labor Wages	477.96	**659.77**	**181.81**	**38.04**	913.00	**1248.70**	**335.71**	**36.77**	34.45	**68.78**	**34.33**	**99.65**	1425.40	**1977.25**	**551.85**	**38.72**
Draft Animals	4.09	*4.18*	*0.09*	*2.30*	10.19	**8.39**	**−1.79**	**−17.59**	0.41	**0.30**	**−0.12**	**−28.35**	14.69	**12.87**	**−1.82**	**−12.36**
Machinery	312.98	*317.75*	*4.77*	*1.52*	560.01	*587.57*	*27.56*	*4.92*	24.05	*25.02*	*0.97*	*4.01*	897.04	*930.33*	*33.29*	*3.71*
Seeds Cost	142.57	*146.00*	*3.43*	*2.40*	230.19	*240.33*	*10.14*	*4.40*	16.21	**16.17**	**−0.04**	**−0.25**	388.98	*402.50*	*13.52*	*3.48*
Manure	47.46	*48.05*	*0.59*	*1.25*	111.17	**109.17**	**−2.00**	**−1.80**	9.32	*12.18*	*2.85*	*30.59*	167.96	*169.40*	*1.45*	*0.86*
Fertilizers	261.95	*266.37*	*4.42*	*1.69*	602.30	**576.26**	**−26.03**	**−4.32**	28.28	**27.67**	**−0.62**	**−2.18**	892.53	**870.30**	**−22.23**	**−2.49**
Insecticides	50.70	*50.70*	*0.00*	*0.00*	28.90	*38.41*	*9.51*	*32.90*	5.46	**5.23**	**−0.23**	**−4.22**	85.06	*94.34*	*9.28*	*10.91*
Laser land leveling cost	18.24	**18.24**	**0.00**	**0.00**	23.76	**23.48**	**−0.27**	**−1.16**	0.75	*1.13*	*0.38*	*50.86*	42.75	*42.85*	*0.10*	*0.25*

(continued)

Table 10.3 (continued)

Upper Egypt	Winter cultivation				Summer cultivation				Nili cultivation				Cultivation in Upper Egypt			
	Mean	AWSCS	Change	%	Mean	AWSCS	Change	%	Mean	AWSCS	Change	%	Mean	AWSCS	Change	%
Other Expenses	124.40	*125.53*	*1.13*	*0.91*	323.05	*335.04*	*11.99*	*3.71*	13.07	*14.15*	*1.08*	*8.24*	460.51	*474.71*	*14.20*	*3.08*
Rent	990.04	*1019.77*	*29.7*	*3.00*	1375.05	*1396.20*	*21.15*	*1.54*	66.58	*67.52*	*0.94*	*1.41*	2431.67	*2483.50*	*51.3*	*2.13*
Crop revenue	19876.41	**27255.50**	**7379.10**	**37.12**	8251.47	**10591.09**	**2339.62**	**28.35**	367.62	**760.18**	**392.56**	**106.78**	28495.50	**38606.77**	**10111.3**	**35.48**
Crop profit	17445.97	**22621.35**	**5175.39**	**29.67**	4630.84	**6027.53**	**1396.69**	**30.16**	249.70	**347.56**	**97.86**	**39.19**	22326.51	**28996.44**	**6669.9**	**29.87**
Crop water consumption	2306.31	**1342.60**	**−963.70**	**−41.79**	6338.33	**3817.88**	**−2520.45**	**−39.77**	243.28	**156.20**	**−87.07**	**−35.79**	8887.91	**5316.68**	**−3571.2**	**−40.18**
Kerosene fuel Liter	334.93	**286.83**	**−48.10**	**−14.36**	331.24	**312.39**	**−18.85**	**−5.69**	21.41	**18.28**	**−3.13**	**−14.60**	687.57	**617.50**	**−70.1**	**−10.19**
Energy consumption in cultivation MJ	10224.81	**8724.37**	**−1500.45**	**−14.67**	19645.26	**17134.74**	**−2510.53**	**−12.78**	1147.22	**975.48**	**−171.74**	**−14.97**	31017.30	**26834.58**	**−4182.7**	**−13.49**

Data source (1) MALR (2015) (2) AWSCS model (2015) (3) MWRI (2015)

Table 10.4 Changes in New land area, unit values and aggregate zones flow values from mean (2011/2012–2013/2014) to AWSCS (bold are values that have increased, italic are values that have decreased)

Lower Egypt	Winter cultivation				Summer cultivation				Nili cultivation				Cultivation in Lower Egypt			
	Mean	AWSCS	Change	%	Mean	AWSCS	Change	%	Mean	AWSCS	Change	%	Mean	AWSCS	Change	%
Irrigated area of crop in old land	237.69	**237.69**	**0.00**	**0.00**	207.94	**207.94**	**0.00**	**0.00**	15.93	**15.93**	**0.00**	**0.00**	461.56	**461.56**	**0.00**	**0.00**
Soil type	0.92	*1.18*	*0.26*	*27.71*	0.47	*0.65*	*0.19*	*39.80*	0.08	**0.07**	**−0.01**	**−10.34**	1.47	*1.90*	*0.43*	*29.54*
Main crop yield	7.27	**11.65**	**4.38**	**60.30**	4.53	**7.54**	**3.01**	**66.57**	0.23	**0.60**	**0.37**	**164.05**	12.02	**19.79**	**7.77**	**64.61**
Secondary crop yield	2.06	**2.90**	**0.84**	**40.80**	1.77	**2.22**	**0.45**	**25.49**	0.07	**0.15**	**0.07**	**95.38**	3.90	**5.26**	**1.36**	**34.91**
Main crop price	654.64	**752.77**	**98.13**	**14.99**	591.80	**922.25**	**330.45**	**55.84**	10.78	**54.49**	**43.72**	**405.66**	1257.22	**1729.52**	**472.30**	**37.57**
Secondary crop price	40.64	**41.64**	**1.00**	**2.47**	6.43	**7.29**	**0.86**	**13.33**	0.20	**0.43**	**0.23**	**112.58**	47.27	**49.36**	**2.09**	**4.42**
Total crop production cost	1906.26	*1972.70*	*66.44*	*3.49*	1764.92	*2364.65*	*599.73*	*33.98*	158.58	*171.05*	*12.47*	*7.87*	3829.76	*4508.40*	*678.64*	*17.72*
*Labor Wages	340.14	**586.66**	**246.52**	**72.48**	412.84	**647.24**	**234.40**	**56.78**	16.44	**48.25**	**31.81**	**193.54**	769.41	**1282.14**	**512.73**	**66.64**
Draft Animals	3.52	**2.79**	**−0.72**	**−20.55**	17.55	**6.12**	**−11.43**	**−65.11**	0.16	**0.13**	**−0.03**	**−15.96**	21.22	**9.05**	**−12.18**	**−57.36**
Machinery	245.12	**238.65**	**−6.47**	**−2.64**	234.30	**221.29**	**−13.00**	**−5.55**	15.23	*16.79*	*1.56*	*10.27*	494.65	**476.74**	**−17.91**	**−3.62**
Seeds Cost	108.53	*110.46*	*1.93*	*1.78*	150.05	*151.04*	*0.99*	*0.66*	14.48	**14.21**	**−0.27**	**−1.88**	273.06	*275.72*	*2.65*	*0.97*
Manure	42.71	**40.06**	**−2.65**	**−6.20**	87.33	**78.46**	**−8.87**	**−10.16**	6.64	*7.42*	*0.78*	*11.70*	136.68	**125.94**	**−10.74**	**−7.86**
Fertilizers	221.04	*229.51*	*8.48*	*3.84*	241.23	**223.71**	**−17.51**	**−7.26**	20.05	*20.07*	*0.02*	*0.12*	482.31	**473.30**	**−9.01**	**−1.87**
Insecticides	56.84	*59.91*	*3.07*	*5.40*	64.87	*66.47*	*1.60*	*2.47*	5.38	*5.49*	*0.11*	*2.03*	127.09	*131.87*	*4.78*	*3.76*
Laser land leveling cost	13.85	**13.85**	**0.00**	**0.00**	8.00	**8.00**	**0.00**	**0.00**	0.63	*0.63*	*0.00*	*0.00*	22.47	**22.47**	**0.00**	**0.00**
Other Expenses	100.65	**99.87**	**−0.79**	**−0.78**	135.53	**118.13**	**−17.40**	**−12.84**	9.58	*10.36*	*0.77*	*8.06*	245.76	**228.35**	**−17.41**	**−7.08**
Rent	915.67	**771.88**	**−143.8**	**−15.70**	721.17	**657.05**	**−64.12**	**−8.89**	45.38	*47.70*	*2.32*	*5.11*	1682.22	**1476.64**	**−205.6**	**−12.22**
Crop revenue	8836.68	**13828.61**	**4991.93**	**56.49**	5732.57	**7604.38**	**1871.80**	**32.65**	148.47	**632.07**	**483.60**	**325.72**	14717.72	**22065.05**	**7347.3**	**49.92**
Crop profit	7287.01	**10318.93**	**3031.92**	**41.61**	4240.11	**5320.07**	**1079.95**	**25.47**	90.06	**302.21**	**212.15**	**235.55**	11617.19	**15941.20**	**4324.0**	**37.22**
Crop water consumption	1391.65	**791.62**	**−600.03**	**−43.12**	2309.53	**1262.90**	**−1046.62**	**−45.32**	122.49	**81.71**	**−40.79**	**−33.30**	3823.67	**2136.23**	**−1687.4**	**−44.13**
Kerosene fuel Liter	368.10	**320.04**	**−48.06**	**−13.06**	193.08	**187.55**	**−5.53**	**−2.87**	12.95	**10.06**	**−2.89**	**−22.35**	574.14	**517.65**	**−56.5**	**−9.84**
Energy consumption in cultivation MJ	8009.23	**7017.60**	**−991.64**	**−12.38**	10246.11	**9413.11**	**−833.01**	**−8.13**	743.59	**661.84**	**−81.75**	**−10.99**	18998.93	**17092.54**	**−1906.4**	**−10.03**

Middle Egypt	Winter cultivation				Summer cultivation				Nili cultivation				Cultivation in Middle Egypt			
	Mean	AWSCS	Change	%	Mean	AWSCS	Change	%	Mean	AWSCS	Change	%	Mean	AWSCS	Change	%
Irrigated area of crop in old land	47.81	**47.81**	**0.00**	**0.00**	31.49	**31.49**	**0.00**	**0.00**	4.17	**4.17**	**0.00**	**0.00**	83.47	**83.47**	**0.00**	**0.00**
Soil type	0.20	**0.19**	**−0.01**	**−5.67**	0.12	*0.20*	*0.07*	*58.32*	0.01	*0.01*	*0.01*	*82.13*	0.34	*0.40*	*0.07*	*20.03*
Main crop yield	1.78	**2.16**	**0.38**	**21.08**	0.88	**1.30**	**0.42**	**47.62**	0.11	**0.18**	**0.07**	**63.90**	2.77	**3.64**	**0.87**	**31.22**
Secondary crop yield	0.52	**0.60**	**0.08**	**14.34**	0.32	**0.48**	**0.16**	**49.51**	0.01	**0.01**	**0.00**	**4.39**	0.86	**1.09**	**0.23**	**27.28**

(continued)

Table 10.4 (continued)

Middle Egypt	Winter cultivation				Summer cultivation				Nili cultivation				Cultivation in Middle Egypt			
	Mean	AWSCS	Change	%	Mean	AWSCS	Change	%	Mean	AWSCS	Change	%	Mean	AWSCS	Change	%
Main crop price	121.94	**174.28**	**52.35**	**42.93**	55.80	**96.45**	**40.64**	**72.83**	6.14	**14.06**	**7.92**	**128.97**	183.88	**284.79**	**100.91**	**54.88**
Secondary crop price	6.31	**7.47**	**1.16**	**18.37**	1.17	**1.51**	**0.34**	**29.05**	0.04	**0.04**	**0.01**	**14.18**	7.52	**9.03**	**1.51**	**20.01**
Total crop production cost	467.19	*473.21*	*6.03*	*1.29*	336.93	**285.78**	**−51.15**	**−15.18**	41.74	*45.52*	*3.88*	*9.29*	845.86	**804.62**	**−41.24**	**−4.88**
*Labor Wages	99.11	**126.87**	**27.77**	**28.02**	62.68	**95.73**	**33.05**	**52.74**	7.08	**12.47**	**5.39**	**76.06**	168.86	**235.07**	**66.21**	**39.21**
Draft Animals	0.59	**0.43**	**−0.15**	**−26.29**	0.60	**0.36**	**−0.24**	**−39.59**	0.02	*0.02*	*0.00*	*3.65*	1.20	**0.81**	**−0.39**	**−32.44**
Machinery	63.57	**51.17**	**−12.40**	**−19.50**	38.92	**29.39**	**−9.53**	**−24.49**	2.68	*4.35*	*1.67*	*62.16*	105.17	**84.91**	**−20.26**	**−19.26**
Seeds Cost	37.63	**32.28**	**−5.35**	**−14.22**	22.86	**18.16**	**−4.70**	**−20.55**	3.26	*3.58*	*0.32*	*9.71*	63.76	**54.02**	**−9.73**	**−15.27**
Manure	13.37	**10.31**	**−3.06**	**−22.87**	13.63	**12.30**	**−1.33**	**−9.75**	0.95	*1.85*	*0.91*	*95.30*	27.95	**24.47**	**−3.48**	**−12.46**
Fertilizers	52.48	**49.71**	**−2.77**	**−5.27**	36.52	**33.26**	**−3.26**	**−8.93**	3.62	*5.08*	*1.46*	*40.35*	92.62	**88.05**	**−4.57**	**−4.93**
Insecticides	13.59	**12.40**	**−1.19**	**−8.76**	8.18	**7.37**	**−0.81**	**−9.94**	1.16	*1.89*	*0.73*	*62.49*	22.94	**21.66**	**−1.28**	**−5.57**
Laser land leveling cost	2.60	**2.60**	**0.00**	**0.00**	1.31	**1.31**	**0.00**	**0.00**	0.12	*0.15*	*0.03*	*27.11*	4.02	*4.06*	*0.03*	*0.81*
Other Expenses	29.05	**23.27**	**−5.78**	**−19.88**	20.21	**16.95**	**−3.26**	**−16.13**	1.73	*2.73*	*1.00*	*57.61*	50.99	**42.95**	**−8.04**	**−15.76**
Rent	155.90	**151.14**	**−4.8**	**−3.05**	101.54	**90.64**	**−10.91**	**−10.74**	7.94	*13.49*	*5.55*	*69.88*	265.38	**255.27**	**−10.1**	**−3.81**
Crop revenue	2356.13	**2843.79**	**487.66**	**20.70**	622.10	**1052.36**	**430.26**	**69.16**	105.62	**183.34**	**77.72**	**73.58**	3083.85	**4079.49**	**995.6**	**32.29**
Crop profit	1923.71	**2331.85**	**408.14**	**21.22**	430.46	**726.82**	**296.36**	**68.85**	67.67	**137.86**	**70.18**	**103.71**	2421.84	**3196.53**	**774.7**	**31.99**
Crop water consumption	304.04	**174.23**	**−129.81**	**−42.69**	308.63	**200.96**	**−107.67**	**−34.89**	36.09	**25.06**	**−11.03**	**−30.56**	648.76	**400.25**	**−248.5**	**−38.30**
Kerosene fuel Liter	86.99	**75.79**	**−11.19**	**−12.87**	27.85	**26.48**	**−1.36**	**−4.89**	3.82	**3.48**	**−0.34**	**−8.87**	118.65	**105.76**	**−12.9**	**−10.87**
Energy consumption in cultivation MJ	1900.88	**1636.29**	**−264.60**	**−13.92**	1467.82	**1329.59**	**−138.24**	**−9.42**	214.17	**204.38**	**−9.79**	**−4.57**	3582.88	**3170.25**	**−412.6**	**−11.52**
Upper Egypt	**Winter cultivation**				**Summer cultivation**				**Nili cultivation**				**Cultivation in Upper Egypt**			
	Mean	AWSCS	Change	%	Mean	AWSCS	Change	%	Mean	AWSCS	Change	%	Mean	AWSCS	Change	%
Irrigated area of crop in old land	87.60	**87.60**	**0.00**	**0.00**	64.37	**64.37**	**0.00**	**0.00**	6.58	**6.58**	**0.00**	**0.00**	158.56	**158.56**	**0.00**	**0.00**
Soil type	0.43	*0.51*	*0.08*	*18.68*	0.32	**0.31**	**−0.02**	**−4.75**	0.02	*0.04*	*0.02*	*69.41*	0.78	*0.86*	*0.08*	*10.37*
Main crop yield	2.71	**4.10**	**1.38**	**50.99**	2.57	**3.88**	**1.31**	**50.81**	0.09	**0.18**	**0.09**	**99.94**	5.38	**8.16**	**2.78**	**51.72**
Secondary crop yield	1.25	**1.70**	**0.45**	**36.18**	0.37	**0.67**	**0.30**	**81.27**	0.03	**0.11**	**0.07**	**204.99**	1.65	**2.48**	**0.32**	**49.87**
Main crop price	190.91	**200.13**	**9.22**	**4.83**	149.13	**235.57**	**86.44**	**57.96**	8.63	**11.28**	**2.65**	**30.66**	348.67	**446.97**	**98.30**	**28.19**
Secondary crop price	16.90	**20.67**	**3.77**	**22.32**	1.18	**2.00**	**0.82**	**69.45**	0.23	**0.32**	**0.09**	**40.35**	18.30	**22.99**	**4.58**	**25.59**
Total crop production cost	898.91	*969.66*	*70.75*	*7.87*	742.52	*803.37*	*60.85*	*8.20*	37.57	*61.26*	*23.69*	*63.05*	1679.00	*1834.29*	*155.29*	*9.25*
*Labor Wages	151.12	**237.87**	**86.75**	**57.41**	141.78	**207.37**	**65.59**	**46.26**	7.80	**17.35**	**9.55**	**122.48**	300.69	**462.58**	**161.89**	**53.84**

(continued)

Table 10.4 (continued)

Upper Egypt	Winter cultivation				Summer cultivation				Nili cultivation				Cultivation in Upper Egypt			
	Mean	AWSCS	Change	%	Mean	AWSCS	Change	%	Mean	AWSCS	Change	%	Mean	AWSCS	Change	%
Draft Animals	1.45	**1.04**	**−0.41**	**−28.46**	1.13	**0.72**	**−0.41**	**−36.13**	0.07	**0.06**	**0.00**	**−6.71**	2.65	**1.82**	**−0.83**	**−31.19**
Machinery	105.58	**103.48**	**−2.10**	**−1.99**	86.41	**81.51**	**−4.91**	**−5.68**	6.28	*6.38*	*0.10*	*1.56*	198.28	**191.36**	**−6.91**	**−3.49**
Seeds Cost	54.79	*55.99*	*1.20*	*2.19*	70.81	**66.55**	**−4.26**	**−6.02**	4.59	**4.23**	**−0.36**	**−7.92**	130.19	**126.77**	**−3.43**	**−2.63**
Manure	20.15	**18.36**	**−1.79**	**−8.89**	13.88	*15.65*	*1.77*	*12.72*	2.65	*2.98*	*0.33*	*12.39*	36.69	*36.99*	*0.30*	*0.83*
Fertilizers	93.85	*95.97*	*2.12*	*2.26*	94.55	*103.26*	*8.72*	*9.22*	6.65	*6.93*	*0.28*	*4.20*	195.04	*206.16*	*11.12*	*5.70*
Insecticides	23.86	**22.40**	**−1.46**	**−6.12**	7.98	*8.60*	*0.63*	*7.84*	1.27	*1.58*	*0.31*	*24.31*	33.11	**32.58**	**−0.53**	**−1.59**
Laser land leveling cost	5.13	**5.13**	**0.00**	**0.00**	3.93	**3.93**	**0.00**	**0.00**	0.20	*0.30*	*0.10*	*48.97*	9.26	*9.36*	*0.10*	*1.06*
Other Expenses	46.45	**44.77**	**−1.67**	**−3.60**	55.66	**53.41**	**−2.25**	**−4.03**	2.08	*3.61*	*1.52*	*73.14*	104.19	**101.79**	**−2.40**	**−2.30**
Rent	332.97	**299.69**	**−33.3**	**−9.99**	205.82	*217.82*	*11.99*	*5.83*	10.53	*17.85*	*7.31*	*69.43*	549.32	**535.36**	**−14.0**	**−2.54**
Crop revenue	2714.06	**3605.48**	**891.42**	**32.84**	1470.72	**2565.11**	**1094.39**	**74.41**	86.49	**126.88**	**40.39**	**46.70**	4271.26	**6297.47**	**2026.2**	**47.44**
Crop profit	2030.77	**2635.81**	**605.05**	**29.79**	958.47	**1823.47**	**865.00**	**90.25**	56.67	**65.93**	**9.26**	**16.34**	3045.91	**4525.22**	**1479.3**	**48.57**
Crop water consumption	656.19	**414.94**	**−241.25**	**−36.76**	1154.48	**725.30**	**−429.18**	**−37.18**	67.94	**44.89**	**−23.05**	**−33.93**	1878.61	**1185.13**	**−693.5**	**−36.91**
Kerosene fuel Liter	131.85	**112.79**	**−19.06**	**−14.45**	63.60	**58.77**	**−4.83**	**−7.59**	5.47	**4.42**	**−1.05**	**−19.26**	200.92	**175.98**	**−24.9**	**−12.41**
Energy consumption in cultivation MJ	3538.66	**3046.23**	**−492.43**	**−13.92**	3545.46	**3194.39**	**−351.07**	**−9.90**	297.11	**266.94**	**−30.17**	**−10.15**	7381.23	**6507.56**	**−873.7**	**−11.84**

Data source (1) MALR (2015) (2) AWSCS model (2015) (3) MWRI (2015)

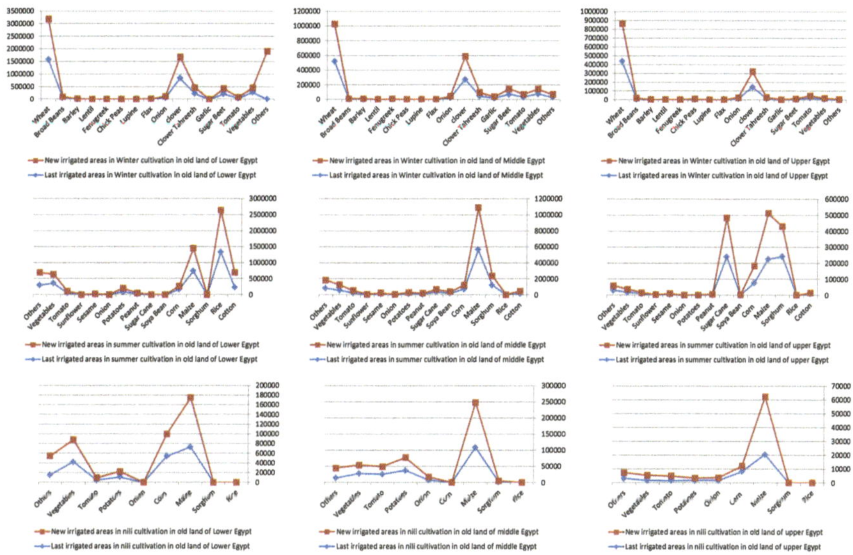

Fig. 10.3 Changes in crops area aggregates in the three zones and seasons flow values from mean (2009/2010–2011/2012) to AWSCS. *Data source* (1) MALR (2015) (2) AWSCS model (2015)

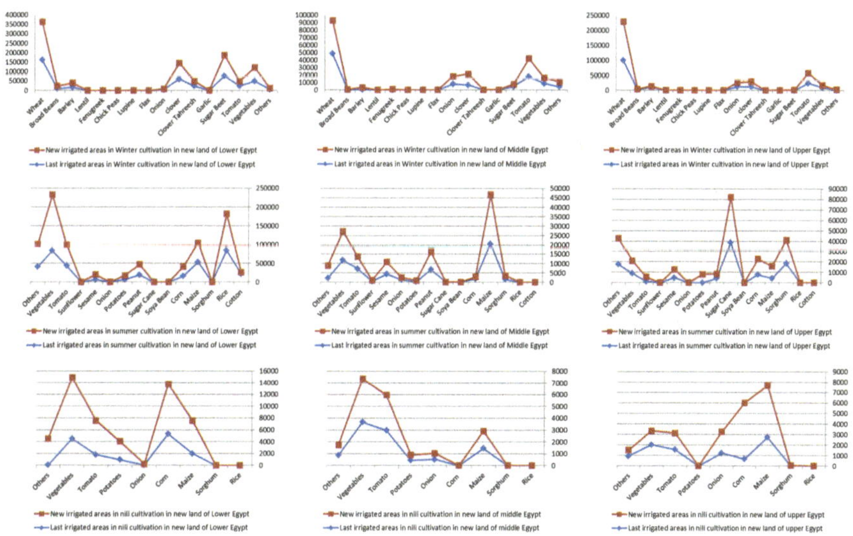

Fig. 10.4 Changes in crops area aggregates in the three zones and seasons flow values from mean (2009/2010–2011/2012) to AWSCS. *Data source* (1) MALR (2015) (2) AWSCS model (2015)

Table 10.5 Changes of economic and financial values in old land zones from mean (2011/2012–2013/2014) to AWSCS (bold are values that have increased, italic are values that have decreased)

Lower Egypt	Winter cultivation				Summer cultivation				Nili cultivation				Cultivation in Lower Egypt			
	Mean	AWSCS	Change	%	Mean	AWSCS	Change	%	Mean	AWSCS	Change	%	Mean	AWSCS	Change	%
Irrigated area of crop in old land	2922.5	**2922.5**	**0.0**	**0.0**	2253.2	2253.2	0.0	0.0	243.5	**243.5**	**0.0**	**0.0**	5419.3	**5419.3**	**0.0**	**0.0**
Total crop production cost	−23410.9	*−26607.4*	*−3196.4*	*13.7*	−24969.4	*−26684.4*	*−1715.0*	*6.9*	−2381.4	*−2484.7*	*−103.4*	*4.3*	−50761.8	**−55776.5**	**−5014.8**	**9.9**
Crop revenue	157470.7	**257548.6**	**100078.0**	**63.6**	57960.8	**76210.0**	**18249.2**	**31.5**	4664.4	**8856.0**	**4191.7**	**89.9**	220095.9	**342614.7**	**122518.8**	**55.7**
Crop profit	139919.9	**226132.7**	**86212.9**	**61.6**	36205.9	**49525.6**	**13319.8**	**36.8**	3142.7	**4370.2**	**1227.4**	**39.1**	179268.5	**280028.5**	**100760.1**	**56.2**
Rate of return (IRR)	5.7	**8.7**	**3.0**	**51.6**	1.3	**1.9**	**0.5**	**40.5**	1.0	**2.6**	**1.6**	**167.5**	3.3	**5.1**	**1.8**	**54.2**
Absolute Risk	0.4	**0.3**	**−0.2**	**−38.9**	0.2	**0.2**	**−0.1**	**−23.9**	0.6	**0.3**	**−0.3**	**−47.3**	0.4	**0.3**	**−0.1**	**−35.8**
Middle Egypt	Winter cultivation				Summer cultivation				Nili cultivation				Cultivation in Middle Egypt			
	Mean	AWSCS	Change	%	Mean	AWSCS	Change	%	Mean	AWSCS	Change	%	Mean	AWSCS	Change	%
Irrigated area of crop in old land	473.5	**473.5**	**0.0**	**0.0**	423.8	**423.8**	**0.0**	**0.0**	113.8	**113.8**	**0.0**	**0.0**	1011.1	**1011.1**	**0.0**	**0.0**
Total crop production cost	−3866.1	*−4731.8*	*−865.7*	*22.4*	−4536.6	*−4584.1*	*−47.5*	*1.0*	−1163.7	*−1234.1*	*−70.5*	*6.1*	−9566.4	**−10550.1**	**−983.7**	**10.3**
Crop revenue	30891.9	**38839.1**	**7947.2**	**25.7**	10071.2	**13621.4**	**3550.2**	**35.3**	2545.3	**4337.2**	**1792.0**	**70.4**	43508.3	**56797.7**	**13289.3**	**30.5**
Crop profit	27025.1	**33748.8**	**6723.7**	**24.9**	6582.9	**9037.3**	**2454.4**	**37.3**	1903.6	**2346.4**	**442.7**	**23.3**	35511.6	**45132.4**	**9620.8**	**27.1**
Rate of return (IRR)	7.0	**7.2**	**0.2**	**3.1**	1.2	**2.0**	**0.8**	**61.6**	1.2	**2.5**	**1.3**	**111.8**	3.5	**4.4**	**0.8**	**23.6**
Absolute Risk	0.2	**0.1**	**0.0**	**−20.5**	0.2	**0.2**	**−0.1**	**−26.1**	0.5	**0.3**	**−0.2**	**−41.3**	0.2	**0.2**	**−0.1**	**−23.4**
Upper Egypt	Winter cultivation				Summer cultivation				Nili cultivation				Cultivation in Upper Egypt			
	Mean	AWSCS	Change	%	Mean	AWSCS	Change	%	Mean	AWSCS	Change	%	Mean	AWSCS	Change	%
Irrigated area of crop in old land	290.0	**290.0**	**0.0**	**0.0**	372.6	**372.6**	**0.0**	**0.0**	25.3	**25.3**	**0.0**	**0.0**	688.0	**688.0**	**0.0**	**0.0**
Total crop production cost	−2477.1	*−2992.2*	*−515.2*	*20.8*	−4388.5	*−4563.6*	*−175.1*	*4.0*	−227.2	*−238.1*	*−11.0*	*4.8*	−7092.7	**−7793.9**	**−701.2**	**9.9**
Crop revenue	19876.4	**27255.5**	**7379.1**	**37.1**	8251.5	**10591.1**	**2339.6**	**28.4**	367.6	**760.2**	**392.6**	**106.8**	28495.5	**38606.8**	**10111.3**	**35.5**
Crop profit	17446.0	**22621.4**	**5175.4**	**29.7**	4630.8	**6027.5**	**1396.7**	**30.2**	249.7	**347.6**	**97.9**	**39.2**	22326.5	**28996.4**	**6669.9**	**29.9**
Rate of return (IRR)	7.0	**8.1**	**1.1**	**15.4**	0.9	**1.3**	**0.4**	**50.0**	0.6	**2.2**	**1.6**	**254.5**	3.0	**4.0**	**0.9**	**31.0**
Absolute Risk	0.3	**0.2**	**−0.1**	**−27.1**	0.2	**0.2**	**0.0**	**−22.1**	0.8	**0.4**	**−0.4**	**−51.6**	0.3	**0.2**	**−0.1**	**−26.2**

Data source (1) MALR (2015) (2) AWSCS model (2015) (3) MWRI (2015)

Table 10.6 Changes of economic and financial values in new land zones from mean (2011/2012–2013/2014) to AWSCS (bold are values that have increased, italic are values that have decreased)

Lower Egypt	Winter cultivation				Summer cultivation				Nili cultivation				Cultivation in Lower Egypt			
	Mean	AWSCS	Change	%	Mean	AWSCS	Change	%	Mean	AWSCS	Change	%	Mean	AWSCS	Change	%
Irrigated area of crop in old land	237.687	**237.687**	**0.000**	**0.000**	207.94	**207.94**	**0.00**	**0.00**	15.93	**15.93**	**0.00**	**0.00**	461.56	**461.56**	**0.00**	**0.00**
Total crop production cost	−1906.256	*−1972.698*	*−66.442*	*3.485*	−1764.92	*−2364.65*	*−599.73*	*33.98*	−158.58	*−171.05*	*−12.47*	*7.87*	−3829.76	*−4508.40*	*−678.64*	*17.72*
Crop revenue	8836.677	**13828.607**	**4991.929**	**56.491**	5732.57	**7604.38**	**1871.80**	**32.65**	148.47	**632.07**	**483.60**	**325.72**	14717.72	**22065.05**	**7347.33**	**49.92**
Crop profit	7287.009	**10318.926**	**3031.916**	**41.607**	4240.11	**5320.07**	**1079.95**	**25.47**	90.06	**302.21**	**212.15**	**235.55**	11617.19	**15941.20**	**4324.02**	**37.22**
Rate of return (IRR)	3.636	**6.010**	**2.374**	**65.309**	2.25	*2.22*	*−0.03*	*−1.43*	−0.06	**2.70**	**2.76**	**−4329.64**	2.84	**3.89**	**1.05**	**36.98**
Absolute Risk	0.399	**0.255**	**−0.144**	**−36.099**	23.09%	**17.41%**	**−5.68%**	**−24.61**	230.32%	**54.10%**	**−176.22%**	**−76.51**	35.30%	**23.55%**	**−11.75%**	**−33.30**

Middle Egypt	Winter cultivation				Summer cultivation				Nili cultivation				Cultivation in Middle Egypt			
	Mean	AWSCS	Change	%	Mean	AWSCS	Change	%	Mean	AWSCS	Change	%	Mean	AWSCS	Change	%
Irrigated area of crop in old land	47.81	**47.81**	**0.00**	**0.00**	31.49	**31.49**	**0.00**	**0.00**	4.17	**4.17**	**0.00**	**0.00**	83.47	**83.47**	**0.00**	**0.00**
Total crop production cost	−467.19	**−473.21**	**−6.03**	**1.29**	−336.93	**−285.78**	**51.15**	**−15.18**	−41.74	**−45.62**	**−3.88**	**9.29**	−845.86	**−804.62**	**41.24**	**−4.88**
Crop revenue	2356.13	**2843.79**	**487.66**	**20.70**	622.10	**1052.36**	**430.26**	**69.16**	105.62	**183.34**	**77.72**	**73.58**	3083.85	**4079.49**	**995.64**	**32.29**
Crop profit	1923.71	**2331.85**	**408.14**	**21.22**	430.46	**726.82**	**296.36**	**68.85**	67.67	**137.86**	**70.18**	**103.71**	2421.84	**3196.53**	**774.69**	**31.99**
Rate of return (IRR)	4.04	**5.01**	**0.97**	**23.90**	0.85	**2.68**	**1.84**	**216.92**	1.53	**3.02**	**1.49**	**97.28**	2.65	**4.07**	**1.42**	**53.83**
Absolute Risk	14.64%	**12.13%**	**−2.51%**	**−17.15**	48.90%	**28.91%**	**−19.99%**	**−40.89**	52.03%	**29.98%**	**−22.06%**	**−42.39**	22.83%	**17.26%**	**−5.57%**	**−24.41**

Upper Egypt	Winter cultivation				Summer cultivation				Nili cultivation				Cultivation in Upper Egypt			
	Mean	AWSCS	Change	%	Mean	AWSCS	Change	%	Mean	AWSCS	Change	%	Mean	AWSCS	Change	%
Irrigated area of crop in old land	87.60	87.60	0.00	0.00	64.37	64.37	0.00	0.00	6.58	6.58	0.00	0.00	158.56	158.56	0.00	0.00
Total crop production cost	−898.91	−969.66	−70.75	7.87	−742.52	−803.37	−60.85	8.20	−37.57	−61.26	−23.69	63.05	−1679.00	−1834.29	−155.29	9.25
Crop revenue	2714.06	3605.48	891.42	32.84	1470.72	2565.11	1094.39	74.41	86.49	126.88	40.39	46.70	4271.26	6297.47	2026.20	47.44
Crop profit	2030.77	2635.81	605.05	29.79	958.47	1823.47	865.00	90.25	56.67	65.93	9.26	16.34	3045.91	4525.22	1479.31	48.57
Rate of return (IRR)	2.02	2.72	0.70	34.62	0.98	2.19	1.21	123.61	1.30	1.07	−0.23	−17.73	1.54	2.43	0.89	57.60
Absolute Risk	23.22%	17.48%	−5.74%	−24.72	52.62%	30.17%	−22.45%	−42.66	33.02%	22.51%	−10.51%	−31.83	33.54%	22.75%	−10.79%	−32.17

Data source (1) MALR (2015) (2) AWSCS model (2015) (3) MWRI (2015)

Table 10.7 Changes crop emission in cultivation in old land zones flow values from mean (2011/2012–2013/2014) to AWSCS (bold is values that have increased, italic are values that have decreased)

Lower Egypt	Winter cultivation				Summer cultivation				Nili cultivation				Cultivation in Lower Egypt			
	Mean	AWSCS	Change	%	Mean	AWSCS	Change	%	Mean	AWSCS	Change	%	Mean	AWSCS	Change	%
NO_x	1.711	1.412	−0.299	−17.491	0.643	0.559	−0.084	−13.101	0.041	0.035	−0.005	−13.393	2.395	2.006	−0.389	−16.241
SO_2	8.255	6.811	−1.444	−17.491	3.105	2.698	−0.407	−13.101	0.198	0.171	−0.026	−13.393	11.557	9.680	−1.877	−16.241
CO_2	8298.159	6846.747	−1451.41	−17.491	3120.892	2712.030	−408.86	−13.101	198.618	172.018	−26.60	−13.393	#######	9730.795	−1886.87	−16.241
SO_3	Nugatory	Nugatory	Nugatory	Nugatory	Nugatory	Nugatory	Nugatory	Nugatory	Nugatory	Nugatory	Nugatory	Nugatory	Nugatory	Nugatory	Nugatory	Nugatory
CO	2.637	2.176	−0.461	−17.491	0.992	0.862	−0.130	−13.101	0.063	0.055	−0.008	−13.393	3.692	3.092	−0.600	−16.241
CH	Nugatory	Nugatory	Nugatory	Nugatory	Nugatory	Nugatory	Nugatory	Nugatory	Nugatory	Nugatory	Nugatory	Nugatory	Nugatory	Nugatory	Nugatory	Nugatory
SPM	Nugatory	Nugatory	Nugatory	Nugatory	Nugatory	Nugatory	Nugatory	Nugatory	Nugatory	Nugatory	Nugatory	Nugatory	Nugatory	Nugatory	Nugatory	Nugatory
Middle Egypt	Winter cultivation				Summer cultivation				Nili cultivation				Cultivation in Middle Egypt			
	Mean	AWSCS	Change	%	Mean	AWSCS	Change	%	Mean	AWSCS	Change	%	Mean	AWSCS	Change	%
NO_x	0.325	0.280	−0.044	−13.686	0.189	0.152	−0.038	−19.874	0.047	0.042	−0.005	−11.320	0.561	0.474	−0.087	−15.574
SO_2	1.566	1.352	−0.214	−13.686	0.914	0.732	−0.182	−19.874	0.229	0.203	−0.026	−11.320	2.709	2.287	−0.422	−15.574
CO_2	1574.494	1359.015	−215.48	−13.686	918.878	736.261	−182.62	−19.874	229.837	203.820	−26.02	−11.320	2723.209	2299.097	−424.11	−15.574
SO_3	Nugatory	Nugatory	Nugatory	Nugatory	Nugatory	Nugatory	Nugatory	Nugatory	Nugatory	Nugatory	Nugatory	Nugatory	Nugatory	Nugatory	Nugatory	Nugatory
CO	0.500	0.432	−0.068	−13.686	0.292	0.234	−0.058	−19.874	0.073	0.065	−0.008	−11.320	0.865	0.731	−0.135	−15.574
CH	Nugatory	Nugatory	Nugatory	Nugatory	Nugatory	Nugatory	Nugatory	Nugatory	Nugatory	Nugatory	Nugatory	Nugatory	Nugatory	Nugatory	Nugatory	Nugatory
SPM	Nugatory	Nugatory	Nugatory	Nugatory	Nugatory	Nugatory	Nugatory	Nugatory	Nugatory	Nugatory	Nugatory	Nugatory	Nugatory	Nugatory	Nugatory	Nugatory
Upper Egypt	Winter cultivation				Summer cultivation				Nili cultivation				Cultivation in Upper Egypt			
	Mean	AWSCS	Change	%	Mean	AWSCS	Change	%	Mean	AWSCS	Change	%	Mean	AWSCS	Change	%
NO_x	0.167	0.143	−0.024	−14.360	0.165	0.156	−0.009	−5.690	0.011	0.009	−0.002	−14.601	0.342	0.308	−0.035	−10.191
SO_2	0.805	0.689	−0.116	−14.360	0.796	0.751	−0.045	−5.690	0.051	0.044	−0.008	−14.601	1.652	1.484	−0.168	−10.191
CO_2	809.050	692.869	−116.18	−14.360	800.139	754.612	−45.53	−5.690	51.713	44.163	−7.55	−14.601	1660.902	1491.643	−169.26	−10.191
SO_3	Nugatory	Nugatory	Nugatory	Nugatory	Nugatory	Nugatory	Nugatory	Nugatory	Nugatory	Nugatory	Nugatory	Nugatory	Nugatory	Nugatory	Nugatory	Nugatory
CO	0.257	0.220	−0.037	−14.360	0.254	0.240	−0.014	−5.690	0.016	0.014	−0.002	−14.601	0.528	0.474	−0.054	−10.191
CH	Nugatory	Nugatory	Nugatory	Nugatory	Nugatory	Nugatory	Nugatory	Nugatory	Nugatory	Nugatory	Nugatory	Nugatory	Nugatory	Nugatory	Nugatory	Nugatory
SPM	Nugatory	Nugatory	Nugatory	Nugatory	Nugatory	Nugatory	Nugatory	Nugatory	Nugatory	Nugatory	Nugatory	Nugatory	Nugatory	Nugatory	Nugatory	Nugatory

Data source (1) MALR (2015) (2) AWSCS model (2015)

Table 10.8 Changes crop emission in cultivation in new land zones flow values from mean (2011/2012–2013/2014) to AWSCS (bold is values that have increased, italic are values that have decreased)

Lower Egypt	Winter cultivation				Summer cultivation				Nili cultivation				Cultivation in Lower Egypt			
	Mean	AWSCS	Change	%	Mean	AWSCS	Change	%	Mean	AWSCS	Change	%	Mean	AWSCS	Change	%
NO_x	0.183	0.159	−0.024	−13.057	0.096	0.093	−0.003	−2.866	0.006	0.005	−0.001	−22.345	0.286	0.258	−0.028	−9.840
SO_2	0.885	0.769	−0.115	−13.057	0.464	0.451	−0.013	−2.866	0.031	0.024	−0.007	−22.345	1.380	1.244	−0.136	−9.840
CO_2	889.194	773.089	−116.10	−13.057	466.416	453.047	−13.37	−2.866	31.292	24.300	−6.99	−22.345	1386.902	1250.436	−136.47	−9.840
SO_3	Nugatory	Nugatory	Nugatory	Nugatory	Nugatory	Nugatory	Nugatory	Nugatory	Nugatory	Nugatory	Nugatory	Nugatory	Nugatory	Nugatory	Nugatory	Nugatory
CO	0.283	0.246	−0.037	−13.057	0.148	0.144	−0.004	−2.866	0.010	0.008	−0.002	−22.345	0.441	0.397	−0.043	−9.840
CH	Nugatory	Nugatory	Nugatory	Nugatory	Nugatory	Nugatory	Nugatory	Nugatory	Nugatory	Nugatory	Nugatory	Nugatory	Nugatory	Nugatory	Nugatory	Nugatory
SPM	Nugatory	Nugatory	Nugatory	Nugatory	Nugatory	Nugatory	Nugatory	Nugatory	Nugatory	Nugatory	Nugatory	Nugatory	Nugatory	Nugatory	Nugatory	Nugatory
Middle Egypt	Winter cultivation				Summer cultivation				Nili cultivation				Cultivation in Middle Egypt			
	Mean	AWSCS	Change	%	Mean	AWSCS	Change	%	Mean	AWSCS	Change	%	Mean	AWSCS	Change	%
NO_x	0.043	0.038	−0.006	−12.868	0.014	0.013	−0.001	−4.894	0.002	0.002	0.000	−8.867	0.059	0.053	−0.006	−10.868
SO_2	0.209	0.182	−0.027	−12.868	0.067	0.064	−0.003	−4.894	0.009	0.008	−0.001	−8.867	0.285	0.254	−0.031	−10.868
CO_2	210.129	183.090	−27.04	−12.868	67.263	63.971	−3.29	−4.894	9.221	8.404	−0.82	−8.867	286.613	255.465	−31.15	−10.868
SO_3	Nugatory	Nugatory	Nugatory	Nugatory	Nugatory	Nugatory	Nugatory	Nugatory	Nugatory	Nugatory	Nugatory	Nugatory	Nugatory	Nugatory	Nugatory	Nugatory
CO	0.067	0.058	−0.009	−12.868	0.021	0.020	−0.001	−4.894	0.003	0.003	0.000	−8.867	0.091	0.081	−0.010	−10.868
CH	Nugatory	Nugatory	Nugatory	Nugatory	Nugatory	Nugatory	Nugatory	Nugatory	Nugatory	Nugatory	Nugatory	Nugatory	Nugatory	Nugatory	Nugatory	Nugatory
SPM	Nugatory	Nugatory	Nugatory	Nugatory	Nugatory	Nugatory	Nugatory	Nugatory	Nugatory	Nugatory	Nugatory	Nugatory	Nugatory	Nugatory	Nugatory	Nugatory
Upper Egypt	Winter cultivation				Summer cultivation				Nili cultivation				Cultivation in Upper Egypt			
	Mean	AWSCS	Change	%	Mean	AWSCS	Change	%	Mean	AWSCS	Change	%	Mean	AWSCS	Change	%
NO_x	0.066	0.056	−0.009	−14.454	0.032	0.029	−0.002	−7.594	0.003	0.002	−0.001	−19.260	0.100	0.088	−0.012	−12.414
SO_2	0.317	0.271	−0.046	−14.454	0.153	0.141	−0.012	−7.594	0.013	0.011	−0.003	−19.260	0.483	0.423	−0.060	−12.414
CO_2	318.491	272.455	−46.04	−14.454	153.644	141.976	−11.67	−7.594	13.215	10.670	−2.55	−19.260	485.350	425.101	−60.25	−12.414
SO_3	Nugatory	Nugatory	Nugatory	Nugatory	Nugatory	Nugatory	Nugatory	Nugatory	Nugatory	Nugatory	Nugatory	Nugatory	Nugatory	Nugatory	Nugatory	Nugatory
CO	0.101	0.087	−0.015	−14.454	0.049	0.045	−0.004	−7.594	0.004	0.003	−0.001	−19.260	0.154	0.135	−0.019	−12.414
CH	Nugatory	Nugatory	Nugatory	Nugatory	Nugatory	Nugatory	Nugatory	Nugatory	Nugatory	Nugatory	Nugatory	Nugatory	Nugatory	Nugatory	Nugatory	Nugatory
SPM	Nugatory	Nugatory	Nugatory	Nugatory	Nugatory	Nugatory	Nugatory	Nugatory	Nugatory	Nugatory	Nugatory	Nugatory	Nugatory	Nugatory	Nugatory	Nugatory

Data source (1) MALR (2015) (2) AWSCS model (2015)

exceed costs. According the financial and economic analyses, the annual internal rate of return (**IRR**) became higher than the existing model for the three zones and increased by 1.807, 0.836, and 0.936% in Lower, Middle, and Upper of old land of Egypt and increased by 1.051, 1.424, and 0.889% in Lower, Middle, and Upper of new land of Egypt, respectively. Absolute risk of optimum cultivation reduced by −14.076, −5.054, −6.571% in Lower, Middle, and Upper of old land of Egypt and reduced by −11.754, −5.572, −10.792% in Lower, Middle, and Upper of new land of Egypt, respectively (Tables 10.5 and 10.6). The proposed model provided low greenhouse gases emission than the existing model for all agriculture operations. Pollutant causes destruction of ecosystem, damage to structures and people's health. The social cost of each ton emission of greenhouse gases and air pollutants was accounted on data of optimal use of energy in old and new land of Egypt in Tables 10.7 and 10.8. Finally laser land leveling should make by the farmers because it is the best solution for the Egyptian question, it is low coast (261.904 Egyptian pounds per hectare in north Egypt, 333.333 Egyptian pounds per hectare in south Egypt), high benefits (61255.438 E.P./hectare), and save water consumption by 38.691%.

References

Abdel-Gawad, S., & Khalil, B. M. (2003). "*Site registry for the nile river water quality monitoring network*". Report No.: WQ-TE-0307-011-FN, National water quality and availability management project, National water quality monitoring component.

Changnon, S. A., & Changnon, D. (2000). Long-term fluctuations in hail incidences in the United States. *Journal of Climate, 13*(3), 658–664.

Dacharry, H. (2000). DEVS based timed hierarchy of formalisms. New York: Academic Press, 2000. https://books.google.com.eg/books?isbn=142007234X.

ECAPMS. (2015). *Egyptian central agency for public mobilization and statistics, selected water related statistics*. http://www.msrintranet.capmas.gov.eg

Forrester, J.W., (1961) Introduction to Coding Theory for Flow Equations of Complex Systems Models. *American Journal of Systems and Software, 2*(6), 146–150.

Forrester, J. W., (1969) Urban Dynamics. M.I.T. Press:Cambridge.

Hamada, Y. M. (2014). *Achieving efficiency and equity in sugar factories and sugar-cropping patterns under climate change in upper and middle Egypt, Agribusiness* (Vol. 30, No. 2, pp. 99–112). Published online in Wiley Online Library (wileyonlinelibrary.com/journal/agr).

Huang, P. (1990). *The peasant family and rural development in the Yangzi delta 1350–1988* (p. 421). Redwood City: Stanford University Press.

Jacobsen, T., Adams, R. M., (1958) Salt and Silt in Ancient Mesopotamian Agriculture: Progressive changes in soil salinity and sedimentation contributed to the breakup of past civilizations. *Science*. 128(3334):1251–8.

MALR. (2015). Ministry of agricultural and land reclamation of Egypt, selected data on costs, prices, and land in production 2014.

Mollard, E. (2001, August 10–12). *The meanders of history: Development and political culture in three deltas (Chao Phraya, Nile and Rhone).* International Water History Association's Conference on The Role of Water in History and Development Bergen, Norway.

MWRI. (2015). *Ministry of water resources and irrigation of Egypt, selected water resources data.* http://www.wldelft.nl/cons/area/rbm/wrp/index

Newell, B., & Wasson, R. (2002). *Social system versus solar system: Why policy makers need history*. Department of Engineering, The Australian National University, Director, Centre for Resource and Environmental Studies, The Australian National University.
Senge, P. M. (1990). The fifth discipline: The art & practice of the learning organization. New York: Doubleday Business.
Smith, D. R. (2000). Information Management for the Intelligent Organization: The Art of Scanning the Environment. *Journal of the Association for Information Science and Technology*, 51(7), 687.
Sweeney, L. B., & Sterman, J. D. (2000). Bathtub Dynamics: Initial Results of a Systems Thinking Inventory. Version 1.2, September 2000. Forthcoming, System Dynamics Review.
Tainter, J.A., (1988). *The Collapse of Complex Societies*. Cambridge: Cambridge University Press.
Thackrah, J. R. (2009). The Rutledge Companion to military conflict since 1945, John Richard Thackrah Whidbey, North Yorkshire August 2008.
Wittfogel, K. A. (1956). *The hydraulic civilizations, Man's role in changing the face of the Earth* (pp. 152–164). Chicago: University of Chicago Press.

FSC
www.fsc.org
MIX
Papier | Fördert gute Waldnutzung
FSC® C083411

Kolibri 360 GmbH
Ferdinand-Jühlke-Straße 7
99095 Erfurt, Deutschland
produktsicherheit@kolibri360.de